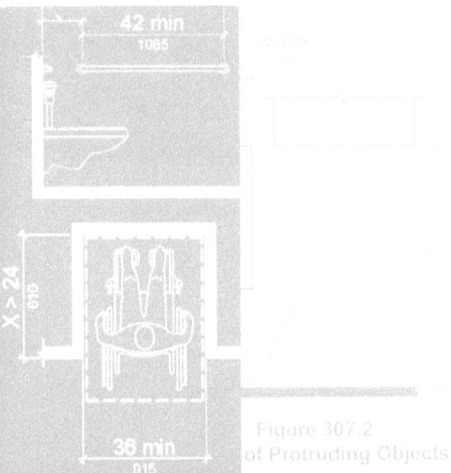

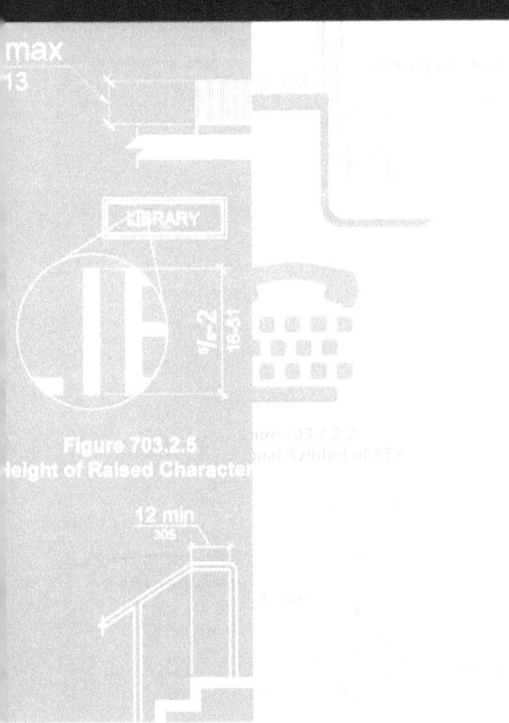

2010 ADA Standards for Accessible Design

Department of Justice
September 15, 2010

Reproduction of this document is encouraged.

This publication may be viewed or downloaded
from the ADA website (www.ADA.gov).
Additional copies may be obtained
by calling the ADA Information Line:

800-514-0301 (voice)
800-514-0383 (TTY)

September 15, 2010

Department of Justice

Contents

1 Introduction..1

2 2010 Standards for State and
Local Government Facilities: Title II....................3

3 2010 Standards for Public Accommodations
and Commercial Facilities: Title III.....................15

4 2010 Standards for Titles II
and III Facilities: 2004 ADAAG............................31

Overview

The Department of Justice published revised regulations for Titles II and III of the Americans with Disabilities Act of 1990 "ADA" in the *Federal Register* on September 15, 2010. These regulations adopted revised, enforceable accessibility standards called the 2010 ADA Standards for Accessible Design "2010 Standards" or "Standards". The 2010 Standards set minimum requirements – both scoping and technical – for newly designed and constructed or altered State and local government facilities, public accommodations, and commercial facilities to be readily accessible to and usable by individuals with disabilities.

Adoption of the 2010 Standards also establishes a revised reference point for Title II entities that choose to make structural changes to existing facilities to meet their program accessibility requirements; and it establishes a similar reference for Title III entities undertaking readily achievable barrier removal.

The Department is providing this document with the official 2010 Standards in one publication. The document includes:

- The 2010 Standards for State and local governments, which consist of the Title II regulations at 28 CFR 35.151 and the 2004 ADAAG at 36 CFR part 1191, appendices B and D;
- The 2010 Standards for public accommodations and commercial facilities, which consist of the Title III regulations at 28 CFR part 36, subpart D, and the 2004 ADAAG at 36 CFR part 1191, appendices B and D.

The Department has assembled into a separate publication the revised regulation guidance that applies to the Standards. The Department included guidance in its revised ADA regulations published on September 15, 2010. This guidance provides detailed information about the Department's adoption of the 2010 Standards including changes to the Standards, the reasoning behind those changes, and responses to public comments received on these topics. The document, Guidance on the 2010 ADA Standards for Accessible Design, can be downloaded from www.ADA.gov.

For More Information

For information about the ADA, including the revised 2010 ADA regulations, please visit the Department's website www.ADA.gov; or, for answers to specific questions, call the toll-free ADA Information Line at 800-514-0301 (Voice) or 800-514-0383 (TTY).

2010 Standards for State and Local Government Facilities: Title II

State and local government facilities must follow the requirements of the 2010 Standards, including both the Title II regulations at 28 CFR 35.151; and the 2004 ADAAG at 36 CFR part 1191, appendices B and D.

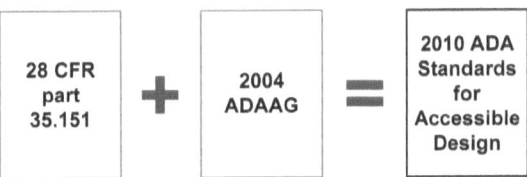

In the few places where requirements between the two differ, the requirements of 28 CFR 35.151 prevail.

Compliance Date for Title II

If the start date for construction is on or after March 15, 2012, all newly constructed or altered State and local government facilities must comply with the 2010 Standards. Before that date, the 1991 Standards (without the elevator exemption), the UFAS, or the 2010 Standards may be used for projects when the start of construction commences on or after September 15, 2010.

Section 35.151 of 28 CFR Part 35

CONTENTS
28 CFR part 35.151 New Construction and Alterations

(a) Design and construction, including the exception for structural impracticability..6

(b) Alterations, including alterations to historic properties, path of travel, and primary function 6

(c) Accessibility standards and compliance date 9

(d) Scope of coverage..11

(e) Social service center establishments...................................11

(f) Housing at a place of education..11

(g) Assembly areas..12

(h) Medical care facilities 13

(i) Curb ramps..13

(j) Facilities with residential dwelling units for sale to individual owners..13

(k) Detention and correctional facilities.....................................13

2004 ADAAG

Chapter 1: Application and Administration.................................37

Chapter 2: Scoping Requirements..50

Chapter 3: Building Blocks..104

Chapter 4: Accessible Routes...117

Chapter 5: General Site and Building Elements.......................149

Chapter 6: Plumbing Elements and Facilities...........................159

Chapter 7: Communication Elements.......................................186

Chapter 8: Special Rooms, Spaces, and Elements..................202

Chapter 9: Built-in Elements...219

Chapter 10: Recreational Facilities...224

Section 35.151 of 28 CFR Part 35

§ 35.151 New construction and alterations.

(a) **Design and construction.**

(1) Each facility or part of a facility constructed by, on behalf of, or for the use of a public entity shall be designed and constructed in such manner that the facility or part of the facility is readily accessible to and usable by individuals with disabilities, if the construction was commenced after January 26, 1992.

(2) **Exception for structural impracticability.**

(i) Full compliance with the requirements of this section is not required where a public entity can demonstrate that it is structurally impracticable to meet the requirements. Full compliance will be considered structurally impracticable only in those rare circumstances when the unique characteristics of terrain prevent the incorporation of accessibility features.

(ii) If full compliance with this section would be structurally impracticable, compliance with this section is required to the extent that it is not structurally impracticable. In that case, any portion of the facility that can be made accessible shall be made accessible to the extent that it is not structurally impracticable.

(iii) If providing accessibility in conformance with this section to individuals with certain disabilities (e.g., those who use wheelchairs) would be structurally impracticable, accessibility shall nonetheless be ensured to persons with other types of disabilities, (e.g., those who use crutches or who have sight, hearing, or mental impairments) in accordance with this section.

(b) **Alterations.**

(1) Each facility or part of a facility altered by, on behalf of, or for the use of a public entity in a manner that affects or could affect the usability of the facility or part of the facility shall, to the maximum extent feasible, be altered in such manner that the altered portion of the facility is readily accessible to and usable by individuals with disabilities, if the alteration was commenced after January 26, 1992.

(2) The path of travel requirements of § 35.151(b)(4) shall apply only to alterations undertaken solely for purposes other than to meet the program accessibility requirements of § 35.150.

(3)

    (i) Alterations to historic properties shall comply, to the maximum extent feasible, with the provisions applicable to historic properties in the design standards specified in § 35.151(c).

    (ii) If it is not feasible to provide physical access to an historic property in a manner that will not threaten or destroy the historic significance of the building or facility, alternative methods of access shall be provided pursuant to the requirements of § 35.150.

(4) Path of travel. An alteration that affects or could affect the usability of or access to an area of a facility that contains a primary function shall be made so as to ensure that, to the maximum extent feasible, the path of travel to the altered area and the restrooms, telephones, and drinking fountains serving the altered area are readily accessible to and usable by individuals with disabilities, including individuals who use wheelchairs, unless the cost and scope of such alterations is disproportionate to the cost of the overall alteration.

    (i) Primary function. A "primary function" is a major activity for which the facility is intended. Areas that contain a primary function include, but are not limited to, the dining area of a cafeteria, the meeting rooms in a conference center, as well as offices and other work areas in which the activities of the public entity using the facility are carried out.

        (A) Mechanical rooms, boiler rooms, supply storage rooms, employee lounges or locker rooms, janitorial closets, entrances, and corridors are not areas containing a primary function. Restrooms are not areas containing a primary function unless the provision of restrooms is a primary purpose of the area, e.g., in highway rest stops.

        (B) For the purposes of this section, alterations to windows, hardware, controls, electrical outlets, and signage shall not be deemed to be alterations that affect the usability of or access to an area containing a primary function.

    (ii) A "path of travel" includes a continuous, unobstructed way of pedestrian passage by means of which the altered area may be approached, entered, and exited, and which connects the altered area with an exterior approach (including sidewalks, streets, and parking areas), an entrance to the facility, and other parts of the facility.

        (A) An accessible path of travel may consist of walks and sidewalks, curb ramps

Section 35.151 of 28 CFR Part 35

and other interior or exterior pedestrian ramps; clear floor paths through lobbies, corridors, rooms, and other improved areas; parking access aisles; elevators and lifts; or a combination of these elements.

(B) For the purposes of this section, the term "path of travel" also includes the restrooms, telephones, and drinking fountains serving the altered area.

(C) **Safe harbor.** If a public entity has constructed or altered required elements of a path of travel in accordance with the specifications in either the 1991 Standards or the Uniform Federal Accessibility Standards before March 15, 2012, the public entity is not required to retrofit such elements to reflect incremental changes in the 2010 Standards solely because of an alteration to a primary function area served by that path of travel.

(iii) **Disproportionality.**

(A) Alterations made to provide an accessible path of travel to the altered area will be deemed disproportionate to the overall alteration when the cost exceeds 20% of the cost of the alteration to the primary function area.

(B) Costs that may be counted as expenditures required to provide an accessible path of travel may include:

(1) Costs associated with providing an accessible entrance and an accessible route to the altered area, for example, the cost of widening doorways or installing ramps;

(2) Costs associated with making restrooms accessible, such as installing grab bars, enlarging toilet stalls, insulating pipes, or installing accessible faucet controls;

(3) Costs associated with providing accessible telephones, such as relocating the telephone to an accessible height, installing amplification devices, or installing a text telephone (TTY); and

(4) Costs associated with relocating an inaccessible drinking fountain.

(iv) **Duty to provide accessible features in the event of disproportionality.**

(A) When the cost of alterations necessary to make the path of travel to the altered area fully accessible is disproportionate to the cost of the overall alteration, the path of travel shall be made accessible to the extent that it can be made accessible without incurring disproportionate costs.

(B) In choosing which accessible elements to provide, priority should be given to those elements that will provide the greatest access, in the following order—

(1) An accessible entrance;
(2) An accessible route to the altered area;
(3) At least one accessible restroom for each sex or a single unisex restroom;
(4) Accessible telephones;
(5) Accessible drinking fountains; and
(6) When possible, additional accessible elements such as parking, storage, and alarms.

(v) Series of smaller alterations.

(A) The obligation to provide an accessible path of travel may not be evaded by performing a series of small alterations to the area served by a single path of travel if those alterations could have been performed as a single undertaking.

(B)

(1) If an area containing a primary function has been altered without providing an accessible path of travel to that area, and subsequent alterations of that area, or a different area on the same path of travel, are undertaken within three years of the original alteration, the total cost of alterations to the primary function areas on that path of travel during the preceding three year period shall be considered in determining whether the cost of making that path of travel accessible is disproportionate.

(2) Only alterations undertaken on or after March 15, 2011, shall be considered in determining if the cost of providing an accessible path of travel is disproportionate to the overall cost of the alterations.

(c) Accessibility standards and compliance date.

(1) If physical construction or alterations commence after July 26, 1992, but prior to the September 15, 2010, then new construction and alterations subject to this section must comply with either the UFAS or the 1991 Standards except that the elevator exemption contained at section 4.1.3(5) and section 4.1.6(1)(k) of the 1991 Standards shall not apply. Departures from particular requirements of either standard by the use of other methods shall be permitted when it is clearly evident that equivalent access to the facility or part of the facility is thereby provided.

Section 35.151 of 28 CFR Part 35

(2) If physical construction or alterations commence on or after September 15, 2010, and before March 15, 2012, then new construction and alterations subject to this section may comply with one of the following: the 2010 Standards, UFAS, or the 1991 Standards except that the elevator exemption contained at section 4.1.3(5) and section 4.1.6(1)(k) of the 1991 Standards shall not apply. Departures from particular requirements of either standard by the use of other methods shall be permitted when it is clearly evident that equivalent access to the facility or part of the facility is thereby provided.

(3) If physical construction or alterations commence on or after March 15, 2012, then new construction and alterations subject to this section shall comply with the 2010 Standards.

(4) For the purposes of this section, ceremonial groundbreaking or razing of structures prior to site preparation do not commence physical construction or alterations.

(5) **Noncomplying new construction and alterations.**

(i) Newly constructed or altered facilities or elements covered by §§ 35.151(a) or (b) that were constructed or altered before March 15, 2012, and that do not comply with the 1991 Standards or with UFAS shall, before March 15, 2012, be made accessible in accordance with either the 1991 Standards, UFAS, or the 2010 Standards.

(ii) Newly constructed or altered facilities or elements covered by §§ 35.151(a) or (b) that were constructed or altered before March 15, 2012 and that do not comply with the 1991 Standards or with UFAS shall, on or after March 15, 2012, be made accessible in accordance with the 2010 Standards.

Appendix to § 35.151(c)

Compliance Date for New Construction or Alterations	Applicable Standards
Before September 15, 2010	1991 Standards or UFAS
On or after September 15, 2010, and before March 15, 2012	1991 Standards, UFAS, or 2010 Standards
On or after March 15, 2012	2010 Standards

Section 35.151 of 28 CFR Part 35

(d) **Scope of coverage.** The 1991 Standards and the 2010 Standards apply to fixed or built-in elements of buildings, structures, site improvements, and pedestrian routes or vehicular ways located on a site. Unless specifically stated otherwise, the advisory notes, appendix notes, and figures contained in the 1991 Standards and the 2010 Standards explain or illustrate the requirements of the rule; they do not establish enforceable requirements.

(e) **Social service center establishments.** Group homes, halfway houses, shelters, or similar social service center establishments that provide either temporary sleeping accommodations or residential dwelling units that are subject to this section shall comply with the provisions of the 2010 Standards applicable to residential facilities, including, but not limited to, the provisions in sections 233 and 809 (pp. 91 and 212).

  (1) In sleeping rooms with more than 25 beds covered by this section, a minimum of 5% of the beds shall have clear floor space complying with section 806.2.3 of the 2010 Standards (p. 209).

  (2) Facilities with more than 50 beds covered by this section that provide common use bathing facilities shall provide at least one roll-in shower with a seat that complies with the relevant provisions of section 608 of the 2010 Standards (p. 174). Transfer-type showers are not permitted in lieu of a roll-in shower with a seat, and the exceptions in sections 608.3 and 608.4 (pp. 177 and 178) for residential dwelling units are not permitted. When separate shower facilities are provided for men and for women, at least one roll-in shower shall be provided for each group.

(f) **Housing at a place of education.** Housing at a place of education that is subject to this section shall comply with the provisions of the 2010 Standards applicable to transient lodging, including, but not limited to, the requirements for transient lodging guest rooms in sections 224 and 806 (pp. 82 and 210) subject to the following exceptions. For the purposes of the application of this section, the term "sleeping room" is intended to be used interchangeably with the term "guest room" as it is used in the transient lodging standards.

  (1) Kitchens within housing units containing accessible sleeping rooms with mobility features (including suites and clustered sleeping rooms) or on floors containing accessible sleeping rooms with mobility features shall provide turning spaces that comply with section 809.2.2 of the 2010 Standards (p. 213) and kitchen work surfaces that comply with section 804.3 of the 2010 Standards (p. 208).

  (2) Multi-bedroom housing units containing accessible sleeping rooms with mobility features shall have an accessible route throughout the unit in accordance with section 809.2 of the 2010 Standards (p. 212).

Section 35.151 of 28 CFR Part 35

(3) Apartments or townhouse facilities that are provided by or on behalf of a place of education, which are leased on a year-round basis exclusively to graduate students or faculty, and do not contain any public use or common use areas available for educational programming, are not subject to the transient lodging standards and shall comply with the requirements for residential facilities in sections 233 and 809 of the 2010 Standards (pp. 91 and 212).

(g) **Assembly areas.** Assembly areas subject to this section shall comply with the provisions of the 2010 Standards applicable to assembly areas, including, but not limited to, sections 221 and 802 (pp. 78 and 202). In addition, assembly areas shall ensure that—

  (1) In stadiums, arenas, and grandstands, wheelchair spaces and companion seats are dispersed to all levels that include seating served by an accessible route;

  (2) Assembly areas that are required to horizontally disperse wheelchair spaces and companion seats by section 221.2.3.1 of the 2010 Standards (p. 79) and have seating encircling, in whole or in part, a field of play or performance area shall disperse wheelchair spaces and companion seats around that field of play or performance area;

  (3) Wheelchair spaces and companion seats are not located on (or obstructed by) temporary platforms or other movable structures, except that when an entire seating section is placed on temporary platforms or other movable structures in an area where fixed seating is not provided, in order to increase seating for an event, wheelchair spaces and companion seats may be placed in that section. When wheelchair spaces and companion seats are not required to accommodate persons eligible for those spaces and seats, individual, removable seats may be placed in those spaces and seats;

  (4) Stadium-style movie theaters shall locate wheelchair spaces and companion seats on a riser or cross-aisle in the stadium section that satisfies at least one of the following criteria—

    (i) It is located within the rear 60% of the seats provided in an auditorium; or

    (ii) It is located within the area of an auditorium in which the vertical viewing angles (as measured to the top of the screen) are from the 40th to the 100th percentile of vertical viewing angles for all seats as ranked from the seats in the first row (1st percentile) to seats in the back row (100th percentile).

Section 35.151 of 28 CFR Part 35

(h) **Medical care facilities.** Medical care facilities that are subject to this section shall comply with the provisions of the 2010 Standards applicable to medical care facilities, including, but not limited to, sections 223 and 805 (pp. 81 and 209). In addition, medical care facilities that do not specialize in the treatment of conditions that affect mobility shall disperse the accessible patient bedrooms required by section 223.2.1 of the 2010 Standards (p. 82) in a manner that is proportionate by type of medical specialty.

(i) **Curb ramps.**

  (1) Newly constructed or altered streets, roads, and highways must contain curb ramps or other sloped areas at any intersection having curbs or other barriers to entry from a street level pedestrian walkway.

  (2) Newly constructed or altered street level pedestrian walkways must contain curb ramps or other sloped areas at intersections to streets, roads, or highways.

(j) **Facilities with residential dwelling units for sale to individual owners.**

  (1) Residential dwelling units designed and constructed or altered by public entities that will be offered for sale to individuals shall comply with the requirements for residential facilities in the 2010 Standards, including sections 233 and 809 (pp. 91 and 212).

  (2) The requirements of paragraph (1) also apply to housing programs that are operated by public entities where design and construction of particular residential dwelling units take place only after a specific buyer has been identified. In such programs, the covered entity must provide the units that comply with the requirements for accessible features to those pre-identified buyers with disabilities who have requested such a unit.

(k) **Detention and correctional facilities.**

  (1) New construction of jails, prisons, and other detention and correctional facilities shall comply with the 2010 Standards except that public entities shall provide accessible mobility features complying with section 807.2 of the 2010 Standards for a minimum of 3%, but no fewer than one, of the total number of cells in a facility (p. 211) Cells with mobility features shall be provided in each classification level.

  (2) **Alterations to detention and correctional facilities.** Alterations to jails, prisons, and other detention and correctional facilities shall comply with the 2010 Standards except that public entities shall provide accessible mobility features complying with section 807.2 of the 2010 Standards for a minimum of 3%, but no fewer than one, of the total number of cells being altered until at least 3%, but no fewer than one, of

Section 35.151 of 28 CFR Part 35

the total number of cells in a facility shall provide mobility features complying with section 807.2 (p. 211). Altered cells with mobility features shall be provided in each classification level. However, when alterations are made to specific cells, detention and correctional facility operators may satisfy their obligation to provide the required number of cells with mobility features by providing the required mobility features in substitute cells (cells other than those where alterations are originally planned), provided that each substitute cell—

(i) Is located within the same prison site;

(ii) Is integrated with other cells to the maximum extent feasible;

(iii) Has, at a minimum, equal physical access as the altered cells to areas used by inmates or detainees for visitation, dining, recreation, educational programs, medical services, work programs, religious services, and participation in other programs that the facility offers to inmates or detainees; and

(iv) If it is technically infeasible to locate a substitute cell within the same prison site, a substitute cell must be provided at another prison site within the corrections system.

(3) With respect to medical and long-term care facilities in jails, prisons, and other detention and correctional facilities, public entities shall apply the 2010 Standards technical and scoping requirements for those facilities irrespective of whether those facilities are licensed.

The remaining text of the 2010 Standards for Title II starts on page 31, under the heading 2010 Standards for Titles II and III: 2004 ADAAG.

2010 Standards for Public Accommodations and Commercial Facilities: Title III

Public accommodations and commercial facilities must follow the requirements of the 2010 Standards, including both the Title III regulations at 28 CFR part 36, subpart D; and the 2004 ADAAG at 36 CFR part 1191, appendices B and D.

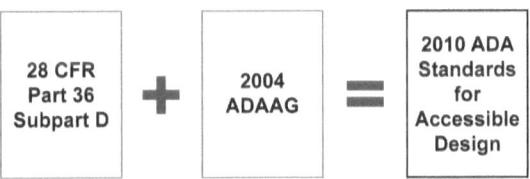

In the few places where requirements between the two differ, the requirements of 28 CFR part 36, subpart D, prevail.

Compliance Date for Title III

The compliance date for the 2010 Standards for new construction and alterations is determined by:

- the date the last application for a building permit or permit extension is certified to be complete by a State, county, or local government;
- the date the last application for a building permit or permit extension is received by a State, county, or local government, where the government does not certify the completion applications; or
- the start of physical construction or alteration, if no permit is required.

If that date is on or after March 15, 2012, then new construction and alterations must comply with the 2010 Standards. If that date is on or after September 15, 2010, and before March 15, 2012, then new construction and alterations must comply with either the 1991 or the 2010 Standards.

CONTENTS

28 CFR part 36, subpart D — New Construction and Alterations

§ Sec.36.401 New construction.
(a) General..19
(b) Commercial facilities located in private residences............................19
(c) Exception for structural impracticability..19
(d) Elevator exemption...20

§ 36.402 Alterations.
(a) General..21
(b) Alteration...21
(c) To the maximum extent feasible 22

§ 36.403 Alterations: Path of travel.
(a) General..22
(b) Primary function..22
(c) Alterations to an area containing a primary function........................22
(d) Landlord/tenant...23
(e) Path of travel...23
(f) Disproportionality...23
(g) Duty to provide accessible features in the
    event of disproportionality..24
(h) Series of smaller alterations..24

§ 36.404 Alterations: Elevator exemption..25

§ 36.405 Alterations: Historic preservation...26

§ 36.406 Standards for new construction and alterations.
(a) Accessibility standards and compliance date..................................26
(b) Scope of coverage..27
(c) Places of lodging...28
(d) Social service center establishments..28
(e) Housing at a place of education..29

Subpart D of 28 CFR Part 36

    (f) Assembly areas..29

    (g) Medical care facilities..30

    § § 36.407—36.499 [Reserved].. 30

2004 ADAAG

    Chapter 1: Application and Administration...37

    Chapter 2: Scoping Requirements.. 50

    Chapter 3: Building Blocks...104

    Chapter 4: Accessible Routes..117

    Chapter 5: General Site and Building Elements.................................149

    Chapter 6: Plumbing Elements and Facilities.....................................159

    Chapter 7: Communication Elements...186

    Chapter 8: Special Rooms, Spaces, and Elements...........................202

    Chapter 9: Built-in Elements..219

    Chapter 10: Recreational Facilities... 224

Subpart D of 28 CFR Part 36

§ 36.401 New construction.

(a) **General.**

    (1) Except as provided in paragraphs (b) and (c) of this section, discrimination for purposes of this part includes a failure to design and construct facilities for first occupancy after January 26, 1993, that are readily accessible to and usable by individuals with disabilities.

    (2) For purposes of this section, a facility is designed and constructed for first occupancy after January 26, 1993, only—

        (i) If the last application for a building permit or permit extension for the facility is certified to be complete, by a State, County, or local government after January 26, 1992 (or, in those jurisdictions where the government does not certify completion of applications, if the last application for a building permit or permit extension for the facility is received by the State, County, or local government after January 26, 1992); and

        (ii) If the first certificate of occupancy for the facility is issued after January 26, 1993.

(b) **Commercial facilities located in private residences.**

    (1) When a commercial facility is located in a private residence, the portion of the residence used exclusively as a residence is not covered by this subpart, but that portion used exclusively in the operation of the commercial facility or that portion used both for the commercial facility and for residential purposes is covered by the new construction and alterations requirements of this subpart.

    (2) The portion of the residence covered under paragraph (b)(1) of this section extends to those elements used to enter the commercial facility, including the homeowner's front sidewalk, if any, the door or entryway, and hallways; and those portions of the residence, interior or exterior, available to or used by employees or visitors of the commercial facility, including restrooms.

(c) **Exception for structural impracticability.**

    (1) Full compliance with the requirements of this section is not required where an entity can demonstrate that it is structurally impracticable to meet the requirements. Full compliance will be considered structurally impracticable only in those rare circumstances when the unique characteristics of terrain prevent the incorporation of accessibility features.

Subpart D of 28 CFR Part 36

(2) If full compliance with this section would be structurally impracticable, compliance with this section is required to the extent that it is not structurally impracticable. In that case, any portion of the facility that can be made accessible shall be made accessible to the extent that it is not structurally impracticable.

(3) If providing accessibility in conformance with this section to individuals with certain disabilities (e.g., those who use wheelchairs) would be structurally impracticable, accessibility shall nonetheless be ensured to persons with other types of disabilities (e.g., those who use crutches or who have sight, hearing, or mental impairments) in accordance with this section.

(d) Elevator exemption.

(1) For purposes of this paragraph (d)—

   (i) **Professional office of a health care provider** means a location where a person or entity regulated by a State to provide professional services related to the physical or mental health of an individual makes such services available to the public. The facility housing the "professional office of a health care provider" only includes floor levels housing at least one health care provider, or any floor level designed or intended for use by at least one health care provider.

   (ii) **Shopping center or shopping mall** means—

   (A) A building housing five or more sales or rental establishments; or

   (B) A series of buildings on a common site, either under common ownership or common control or developed either as one project or as a series of related projects, housing five or more sales or rental establishments. For purposes of this section, places of public accommodation of the types listed in paragraph (5) of the definition of "place of public accommodation" in section § 36.104 are considered sales or rental establishments. The facility housing a "shopping center or shopping mall" only includes floor levels housing at least one sales or rental establishment, or any floor level designed or intended for use by at least one sales or rental establishment.

(2) This section does not require the installation of an elevator in a facility that is less than three stories or has less than 3000 square feet per story, except with respect to any facility that houses one or more of the following:

   (i) A shopping center or shopping mall, or a professional office of a health care provider.

(ii) A terminal, depot, or other station used for specified public transportation, or an airport passenger terminal. In such a facility, any area housing passenger services, including boarding and debarking, loading and unloading, baggage claim, dining facilities, and other common areas open to the public, must be on an accessible route from an accessible entrance.

(3) The elevator exemption set forth in this paragraph (d) does not obviate or limit, in any way the obligation to comply with the other accessibility requirements established in paragraph (a) of this section. For example, in a facility that houses a shopping center or shopping mall, or a professional office of a health care provider, the floors that are above or below an accessible ground floor and that do not house sales or rental establishments or a professional office of a health care provider, must meet the requirements of this section but for the elevator.

§ 36.402 Alterations.

(a) General.

(1) Any alteration to a place of public accommodation or a commercial facility, after January 26, 1992, shall be made so as to ensure that, to the maximum extent feasible, the altered portions of the facility are readily accessible to and usable by individuals with disabilities, including individuals who use wheelchairs.

(2) An alteration is deemed to be undertaken after January 26, 1992, if the physical alteration of the property begins after that date.

(b) Alteration. For the purposes of this part, an alteration is a change to a place of public accommodation or a commercial facility that affects or could affect the usability of the building or facility or any part thereof.

(1) Alterations include, but are not limited to, remodeling, renovation, rehabilitation, reconstruction, historic restoration, changes or rearrangement in structural parts or elements, and changes or rearrangement in the plan configuration of walls and full-height partitions. Normal maintenance, reroofing, painting or wallpapering, asbestos removal, or changes to mechanical and electrical systems are not alterations unless they affect the usability of the building or facility.

(2) If existing elements, spaces, or common areas are altered, then each such altered element, space, or area shall comply with the applicable provisions of appendix A to this part.

Subpart D of 28 CFR Part 36

(c) **To the maximum extent feasible.** The phrase "to the maximum extent feasible," as used in this section, applies to the occasional case where the nature of an existing facility makes it virtually impossible to comply fully with applicable accessibility standards through a planned alteration. In these circumstances, the alteration shall provide the maximum physical accessibility feasible. Any altered features of the facility that can be made accessible shall be made accessible. If providing accessibility in conformance with this section to individuals with certain disabilities (e.g., those who use wheelchairs) would not be feasible, the facility shall be made accessible to persons with other types of disabilities (e.g., those who use crutches, those who have impaired vision or hearing, or those who have other impairments).

§ 36.403 Alterations: Path of travel.

(a) **General.**

  (1) An alteration that affects or could affect the usability of or access to an area of a facility that contains a primary function shall be made so as to ensure that, to the maximum extent feasible, the path of travel to the altered area and the restrooms, telephones, and drinking fountains serving the altered area, are readily accessible to and usable by individuals with disabilities, including individuals who use wheelchairs, unless the cost and scope of such alterations is disproportionate to the cost of the overall alteration.

  (2) If a private entity has constructed or altered required elements of a path of travel at a place of public accommodation or commercial facility in accordance with the specifications in the 1991 Standards, the private entity is not required to retrofit such elements to reflect the incremental changes in the 2010 Standards solely because of an alteration to a primary function area served by that path of travel.

(b) **Primary function.** A "primary function" is a major activity for which the facility is intended. Areas that contain a primary function include, but are not limited to, the customer services lobby of a bank, the dining area of a cafeteria, the meeting rooms in a conference center, as well as offices and other work areas in which the activities of the public accommodation or other private entity using the facility are carried out. Mechanical rooms, boiler rooms, supply storage rooms, employee lounges or locker rooms, janitorial closets, entrances, corridors, and restrooms are not areas containing a primary function.

(c) **Alterations to an area containing a primary function.**

  (1) Alterations that affect the usability of or access to an area containing a primary

function include, but are not limited to—

   (i) Remodeling merchandise display areas or employee work areas in a department store;

   (ii) Replacing an inaccessible floor surface in the customer service or employee work areas of a bank;

   (iii) Redesigning the assembly line area of a factory; or

   (iv) Installing a computer center in an accounting firm.

   (2) For the purposes of this section, alterations to windows, hardware, controls, electrical outlets, and signage shall not be deemed to be alterations that affect the usability of or access to an area containing a primary function.

(d) **Landlord/tenant:** If a tenant is making alterations as defined in § 36.402 that would trigger the requirements of this section, those alterations by the tenant in areas that only the tenant occupies do not trigger a path of travel obligation upon the landlord with respect to areas of the facility under the landlord's authority, if those areas are not otherwise being altered.

(e) **Path of travel.**

   (1) A "path of travel" includes a continuous, unobstructed way of pedestrian passage by means of which the altered area may be approached, entered, and exited, and which connects the altered area with an exterior approach (including sidewalks, streets, and parking areas), an entrance to the facility, and other parts of the facility.

   (2) An accessible path of travel may consist of walks and sidewalks, curb ramps and other interior or exterior pedestrian ramps; clear floor paths through lobbies, corridors, rooms, and other improved areas; parking access aisles; elevators and lifts; or a combination of these elements.

   (3) For the purposes of this part, the term "path of travel" also includes the restrooms, telephones, and drinking fountains serving the altered area.

(f) **Disproportionality.**

   (1) Alterations made to provide an accessible path of travel to the altered area will be deemed disproportionate to the overall alteration when the cost exceeds 20% of the cost of the alteration to the primary function area.

Subpart D of 28 CFR Part 36

(2) Costs that may be counted as expenditures required to provide an accessible path of travel may include:

(i) Costs associated with providing an accessible entrance and an accessible route to the altered area, for example, the cost of widening doorways or installing ramps;

(ii) Costs associated with making restrooms accessible, such as installing grab bars, enlarging toilet stalls, insulating pipes, or installing accessible faucet controls;

(iii) Costs associated with providing accessible telephones, such as relocating the telephone to an accessible height, installing amplification devices, or installing a text telephone (TTY).

(iv) Costs associated with relocating an inaccessible drinking fountain.

(g) Duty to provide accessible features in the event of disproportionality.

(1) When the cost of alterations necessary to make the path of travel to the altered area fully accessible is disproportionate to the cost of the overall alteration, the path of travel shall be made accessible to the extent that it can be made accessible without incurring disproportionate costs.

(2) In choosing which accessible elements to provide, priority should be given to those elements that will provide the greatest access, in the following order:

(i) An accessible entrance;

(ii) An accessible route to the altered area;

(iii) At least one accessible restroom for each sex or a single unisex restroom;

(iv) Accessible telephones;

(v) Accessible drinking fountains; and

(vi) When possible, additional accessible elements such as parking, storage, and alarms.

(h) Series of smaller alterations.

(1) The obligation to provide an accessible path of travel may not be evaded by

performing a series of small alterations to the area served by a single path of travel if those alterations could have been performed as a single undertaking.

(2)

(i) If an area containing a primary function has been altered without providing an accessible path of travel to that area, and subsequent alterations of that area, or a different area on the same path of travel, are undertaken within three years of the original alteration, the total cost of alterations to the primary function areas on that path of travel during the preceding three year period shall be considered in determining whether the cost of making that path of travel accessible is disproportionate.

(ii) Only alterations undertaken after January 26, 1992, shall be considered in determining if the cost of providing an accessible path of travel is disproportionate to the overall cost of the alterations.

§ 36.404 Alterations: Elevator exemption.

(a) This section does not require the installation of an elevator in an altered facility that is less than three stories or has less than 3,000 square feet per story, except with respect to any facility that houses a shopping center, a shopping mall, the professional office of a health care provider, a terminal, depot, or other station used for specified public transportation, or an airport passenger terminal.

(1) For the purposes of this section, professional office of a health care provider means a location where a person or entity regulated by a State to provide professional services related to the physical or mental health of an individual makes such services available to the public. The facility that houses a professional office of a health care provider only includes floor levels housing by at least one health care provider, or any floor level designed or intended for use by at least one health care provider.

(2) For the purposes of this section, shopping center or shopping mall means—

(i) A building housing five or more sales or rental establishments; or

(ii) A series of buildings on a common site, connected by a common pedestrian access route above or below the ground floor, that is either under common ownership or common control or developed either as one project or as a series of related projects, housing five or more sales or rental establishments. For purposes of this section, places of public accommodation of the types listed in paragraph (5) of the definition of place of public accommodation in § 36.104 are considered sales or rental establishments. The facility housing a shopping center or shopping mall only includes floor levels housing at least one sales or rental establishment, or any floor level designed or intended for use by at least one sales or rental establishment.

(b) The exemption provided in paragraph (a) of this section does not obviate or limit in any way the obligation to comply with the other accessibility requirements established in this subpart. For example, alterations to floors above or below the accessible ground floor must be accessible regardless of whether the altered facility has an elevator.

§ 36.405 Alterations: Historic preservation.

(a) Alterations to buildings or facilities that are eligible for listing in the National Register of Historic Places under the National Historic Preservation Act (16 U.S.C. 470 *et seq.*), or are designated as historic under State or local law, shall comply to the maximum extent feasible with this part.

(b) If it is determined that it is not feasible to provide physical access to an historic property that is a place of public accommodation in a manner that will not threaten or destroy the historic significance of the building or the facility, alternative methods of access shall be provided pursuant to the requirements of subpart C of this part.

§ 36.406 Standards for new construction and alterations.

(a) **Accessibility standards and compliance date.**

(1) New construction and alterations subject to §§ 36.401 or 36.402 shall comply with the 1991 Standards if the date when the last application for a building permit or permit extension is certified to be complete by a State, county, or local government (or, in those jurisdictions where the government does not certify completion of applications, if the date when the last application for a building permit or permit extension is received by the State, county, or local government) is before September 15, 2010, or if no permit is required, if the start of physical construction or alterations occurs before September 15, 2010.

(2) New construction and alterations subject to §§ 36.401 or 36.402 shall comply either with the 1991 Standards or with the 2010 Standards if the date when the last application for a building permit or permit extension is certified to be complete by a State, county, or local government (or, in those jurisdictions where the government does not certify completion of applications, if the date when the last application for a building permit or permit extension is received by the State, county, or local government) is on or after September 15, 2010, and before March 15, 2012,

or if no permit is required, if the start of physical construction or alterations occurs on or after September 15, 2010, and before March 15, 2012.

(3) New construction and alterations subject to §§ 36.401 or 36.402 shall comply with the 2010 Standards if the date when the last application for a building permit or permit extension is certified to be complete by a State, county, or local government (or, in those jurisdictions where the government does not certify completion of applications, if the date when the last application for a building permit or permit extension is received by the State, county, or local government) is on or after March 15, 2012, or if no permit is required, if the start of physical construction or alterations occurs on or after March 15, 2012.

(4) For the purposes of this section, "start of physical construction or alterations" does not mean ceremonial groundbreaking or razing of structures prior to site preparation.

(5) **Noncomplying new construction and alterations.**

   (i) Newly constructed or altered facilities or elements covered by §§ 36.401 or 36.402 that were constructed or altered before March 15, 2012, and that do not comply with the 1991 Standards shall, before March 15, 2012, be made accessible in accordance with either the 1991 Standards or the 2010 Standards.

   (ii) Newly constructed or altered facilities or elements covered by §§ 36.401 or 36.402 that were constructed or altered before March 15, 2012 and that do not comply with the 1991 Standards shall, on or after March 15, 2012, be made accessible in accordance with the 2010 Standards.

Appendix to § 36.406(a)

Compliance Dates for New Construction and Alterations	Applicable Standards
On or after January 26, 1993, and before September 15, 2010	1991 Standards
On or after September 15, 2010, and before March 15, 2012	1991 Standards or 2010 Standards
On or after March 15, 2012	2010 Standards

(b) **Scope of coverage.** The 1991 Standards and the 2010 Standards apply to fixed or built-in elements of buildings, structures, site improvements, and pedestrian routes or vehicular ways located on a site. Unless specifically stated otherwise, advisory notes,

Subpart D of 28 CFR Part 36

appendix notes, and figures contained in the 1991 Standards and 2010 Standards explain or illustrate the requirements of the rule; they do not establish enforceable requirements.

(c) **Places of lodging.** Places of lodging subject to this part shall comply with the provisions of the 2010 Standards applicable to transient lodging, including, but not limited to, the requirements for transient lodging guest rooms in sections 224 and 806 of the 2010 Standards (pp. 82 and 210).

  (1) **Guest rooms.** Guest rooms with mobility features in places of lodging subject to the transient lodging requirements of 2010 Standards shall be provided as follows—

   (i) Facilities that are subject to the same permit application on a common site that each have 50 or fewer guest rooms may be combined for the purposes of determining the required number of accessible rooms and type of accessible bathing facility in accordance with table 224.2 to section 224.2 of the 2010 Standards (pp 83).

   (ii) Facilities with more than 50 guest rooms shall be treated separately for the purposes of determining the required number of accessible rooms and type of accessible bathing facility in accordance with table 224.2 to section 224.2 of the 2010 Standards (p. 83).

  (2) **Exception.** Alterations to guest rooms in places of lodging where the guest rooms are not owned or substantially controlled by the entity that owns, leases, or operates the overall facility and the physical features of the guest room interiors are controlled by their individual owners are not required to comply with § 36.402 or the alterations requirements in section 224.1.1 of the 2010 Standards (p. 83).

  (3) **Facilities with residential units and transient lodging units.** Residential dwelling units that are designed and constructed for residential use exclusively are not subject to the transient lodging standards.

(d) **Social service center establishments.** Group homes, halfway houses, shelters, or similar social service center establishments that provide either temporary sleeping accommodations or residential dwelling units that are subject to this part shall comply with the provisions of the 2010 Standards applicable to residential facilities, including, but not limited to, the provisions in sections 233 and 809
(pp. 91 and 212) .

  (1) In sleeping rooms with more than 25 beds covered by this part, a minimum of 5% of the beds shall have clear floor space complying with section 806.2.3 of the 2010 Standards (p. 210).

(2) Facilities with more than 50 beds covered by this part that provide common use bathing facilities shall provide at least one roll-in shower with a seat that complies with the relevant provisions of section 608 of the 2010 Standards (p. 174). Transfer-type showers are not permitted in lieu of a roll-in shower with a seat, and the exceptions in sections 608.3 and 608.4 for residential dwelling units are not permitted. When separate shower facilities are provided for men and for women, at least one roll-in shower shall be provided for each group.

(e) Housing at a place of education. Housing at a place of education that is subject to this part shall comply with the provisions of the 2010 Standards applicable to transient lodging, including, but not limited to, the requirements for transient lodging guest rooms in sections 224 and 806 (pp. 82 and 210), subject to the following exceptions. For the purposes of the application of this section, the term "sleeping room" is intended to be used interchangeably with the term "guest room" as it is used in the transient lodging standards.

(1) Kitchens within housing units containing accessible sleeping rooms with mobility features (including suites and clustered sleeping rooms) or on floors containing accessible sleeping rooms with mobility features shall provide turning spaces that comply with section 809.2.2 of the 2010 Standards (p. 213) and kitchen work surfaces that comply with section 804.3 of the 2010 Standards (p. 208).

(2) Multi-bedroom housing units containing accessible sleeping rooms with mobility features shall have an accessible route throughout the unit in accordance with section 809.2 of the 2010 Standards (p. 212).

(3) Apartments or townhouse facilities that are provided by or on behalf of a place of education, which are leased on a year-round basis exclusively to graduate students or faculty and do not contain any public use or common use areas available for educational programming, are not subject to the transient lodging standards and shall comply with the requirements for residential facilities in sections 233 and 809 of the 2010 Standards (pp. 91 and 212).

(f) Assembly areas. Assembly areas that are subject to this part shall comply with the provisions of the 2010 Standards applicable to assembly areas, including, but not limited to, sections 221 and 802 (p. 78 and 202). In addition, assembly areas shall ensure that—

(1) In stadiums, arenas, and grandstands, wheelchair spaces and companion seats are dispersed to all levels that include seating served by an accessible route;

(2) Assembly areas that are required to horizontally disperse wheelchair spaces and

Subpart D of 28 CFR Part 36

companion seats by section 221.2.3.1 of the 2010 Standards (p. 79) and that have seating encircling, in whole or in part, a field of play or performance, wheelchair spaces and companion seats are dispersed around that field of play or performance area;

(3) Wheelchair spaces and companion seats are not located on (or obstructed by) temporary platforms or other movable structures, except that when an entire seating section is placed on temporary platforms or other movable structures in an area where fixed seating is not provided, in order to increase seating for an event, wheelchair spaces and companion seats may be placed in that section. When wheelchair spaces and companion seats are not required to accommodate persons eligible for those spaces and seats, individual, removable seats may be placed in those spaces and seats;

(4) In stadium-style movie theaters, wheelchair spaces and companion seats are located on a riser or cross-aisle in the stadium section that satisfies at least one of the following criteria—

(i) It is located within the rear 60% of the seats provided in an auditorium; or

(ii) It is located within the area of an auditorium in which the vertical viewing angles (as measured to the top of the screen) are from the 40th to the 100th percentile of vertical viewing angles for all seats as ranked from the seats in the first row (1st percentile) to seats in the back row (100th percentile).

(g) **Medical care facilities.** Medical care facilities that are subject to this part shall comply with the provisions of the 2010 Standards applicable to medical care facilities, including, but not limited to, sections 223 and 805 (pp. 81 and 209). In addition, medical care facilities that do not specialize in the treatment of conditions that affect mobility shall disperse the accessible patient bedrooms required by section 223.2.1 of the 2010 Standards (p. 82) in a manner that is proportionate by type of medical specialty.

§§ 36.407—36.499 [Reserved]

The remaining text of the 2010 Standards for Title III start on page 31, under the heading 2010 Standards for Titles II and III: 2004 ADAAG.

2010 Standards for Titles II and III Facilities: 2004 ADAAG

The following section applies to **both** State and local government facilities (Title II) and public accommodations and commercial facilities (Title III). The section consists of (ADA) Chapters 1 and 2 and Chapters 3 through 10, of the 2004 ADAAG (36 CFR part 1191, appendices B and D, adopted as part of both the Title II and Title III 2010 Standards).

State and local government facilities must follow the requirements of the 2010 Standards, including both the Title II regulations at 28 CFR 35.151; and the 2004 ADAAG at 36 CFR part 1191, appendices B and D.

Public accommodations and commercial facilities must follow the requirements of the 2010 Standards, including both the Title III regulations at 28 CFR part 36, subpart D; and the 2004 ADAAG at 36 CFR part 1191, appendices B and D.

In the few places where requirements between the regulation and the 2004 ADAAG differ, the requirements of 28 CFR 35.151 or 28 CFR part 36, subpart D, prevail.

TABLE OF CONTENTS

ADA CHAPTER 1: APPLICATION AND ADMINISTRATION	**37**
101 Purpose	37
102 Dimensions for Adults and Children	37
103 Equivalent Facilitation	37
104 Conventions	37
105 Referenced Standards	40
106 Definitions	44
ADA CHAPTER 2: SCOPING REQUIREMENTS	**50**
201 Application	50
202 Existing Buildings and Facilities	50
203 General Exceptions	53
204 Protruding Objects	54
205 Operable Parts	55
206 Accessible Routes	55
207 Accessible Means of Egress	64
208 Parking Spaces	65
209 Passenger Loading Zones and Bus Stops	67
210 Stairways	68
211 Drinking Fountains	68
212 Kitchens, Kitchenettes, and Sinks	69
213 Toilet Facilities and Bathing Facilities	69
214 Washing Machines and Clothes Dryers	70
215 Fire Alarm Systems	71
216 Signs	71
217 Telephones	74
218 Transportation Facilities	76
219 Assistive Listening Systems	76
220 Automatic Teller Machines and Fare Machines	77
221 Assembly Areas	78
222 Dressing, Fitting, and Locker Rooms	81
223 Medical Care and Long-Term Care Facilities	81
224 Transient Lodging Guest Rooms	82
225 Storage	85
226 Dining Surfaces and Work Surfaces	86
227 Sales and Service	87
228 Depositories, Vending Machines, Change Machines, Mail Boxes, and Fuel Dispensers	88
229 Windows	88
230 Two-Way Communication Systems	88
231 Judicial Facilities	89
232 Detention Facilities and Correctional Facilities	89
233 Residential Facilities	91
234 Amusement Rides	94
235 Recreational Boating Facilities	95
236 Exercise Machines and Equipment	97
237 Fishing Piers and Platforms	97
238 Golf Facilities	97
239 Miniature Golf Facilities	98

TABLE OF CONTENTS

240 Play Areas	98
241 Saunas and Steam Rooms	102
242 Swimming Pools, Wading Pools, and Spas	102
243 Shooting Facilities with Firing Positions	103

CHAPTER 3: BUILDING BLOCKS — 104

301 General	104
302 Floor or Ground Surfaces	104
303 Changes in Level	105
304 Turning Space	106
305 Clear Floor or Ground Space	107
306 Knee and Toe Clearance	109
307 Protruding Objects	111
308 Reach Ranges	113
309 Operable Parts	116

CHAPTER 4: ACCESSIBLE ROUTES — 117

401 General	117
402 Accessible Routes	117
403 Walking Surfaces	117
404 Doors, Doorways, and Gates	119
405 Ramps	127
406 Curb Ramps	131
407 Elevators	133
408 Limited-Use/Limited-Application Elevators	143
409 Private Residence Elevators	145
410 Platform Lifts	147

CHAPTER 5: GENERAL SITE AND BUILDING ELEMENTS — 149

501 General	149
502 Parking Spaces	149
503 Passenger Loading Zones	152
504 Stairways	153
505 Handrails	154

CHAPTER 6: PLUMBING ELEMENTS AND FACILITIES — 159

601 General	159
602 Drinking Fountains	159
603 Toilet and Bathing Rooms	160
604 Water Closets and Toilet Compartments	161
605 Urinals	170
606 Lavatories and Sinks	170
607 Bathtubs	171
608 Shower Compartments	174
609 Grab Bars	181

TABLE OF CONTENTS

610 Seats	182
611 Washing Machines and Clothes Dryers	185
612 Saunas and Steam Rooms	185

CHAPTER 7: COMMUNICATION ELEMENTS AND FEATURES — 186

701 General	186
702 Fire Alarm Systems	186
703 Signs	186
704 Telephones	194
705 Detectable Warnings	196
706 Assistive Listening Systems	197
707 Automatic Teller Machines and Fare Machines	198
708 Two-Way Communication Systems	201

CHAPTER 8: SPECIAL ROOMS, SPACES, AND ELEMENTS — 202

801 General	202
802 Wheelchair Spaces, Companion Seats, and Designated Aisle Seats	202
803 Dressing, Fitting, and Locker Rooms	206
804 Kitchens and Kitchenettes	206
805 Medical Care and Long-Term Care Facilities	209
806 Transient Lodging Guest Rooms	210
807 Holding Cells and Housing Cells	211
808 Courtrooms	212
809 Residential Dwelling Units	212
810 Transportation Facilities	214
811 Storage	218

CHAPTER 9: BUILT-IN ELEMENTS — 219

901 General	219
902 Dining Surfaces and Work Surfaces	219
903 Benches	219
904 Check-Out Aisles and Sales and Service Counters	220

CHAPTER 10: RECREATION FACILITIES — 224

1001 General	224
1002 Amusement Rides	224
1003 Recreational Boating Facilities	228
1004 Exercise Machines and Equipment	233
1005 Fishing Piers and Platforms	234
1006 Golf Facilities	235
1007 Miniature Golf Facilities	236
1008 Play Areas	237
1009 Swimming Pools, Wading Pools, and Spas	242
1010 Shooting Facilities with Firing Positions	252

INDEX AND LIST OF FIGURES — 253

page intentionally left blank

ADA CHAPTER 1: APPLICATION AND ADMINISTRATION

101 Purpose

101.1 General. This document contains scoping and technical requirements for *accessibility* to *sites, facilities, buildings,* and *elements* by individuals with disabilities. The requirements are to be applied during the design, construction, *additions* to, and *alteration* of *sites, facilities, buildings,* and *elements* to the extent required by regulations issued by Federal agencies under the Americans with Disabilities Act of 1990 (ADA).

> **Advisory 101.1 General.** In addition to these requirements, covered entities must comply with the regulations issued by the Department of Justice and the Department of Transportation under the Americans with Disabilities Act. There are issues affecting individuals with disabilities which are not addressed by these requirements, but which are covered by the Department of Justice and the Department of Transportation regulations.

101.2 Effect on Removal of Barriers in Existing Facilities. This document does not address existing *facilities* unless *altered* at the discretion of a covered entity. The Department of Justice has authority over existing *facilities* that are subject to the requirement for removal of barriers under title III of the ADA. Any determination that this document applies to existing *facilities* subject to the barrier removal requirement is solely within the discretion of the Department of Justice and is effective only to the extent required by regulations issued by the Department of Justice.

102 Dimensions for Adults and Children

The technical requirements are based on adult dimensions and anthropometrics. In addition, this document includes technical requirements based on children's dimensions and anthropometrics for drinking fountains, water closets, toilet compartments, lavatories and sinks, dining surfaces, and work surfaces.

103 Equivalent Facilitation

Nothing in these requirements prevents the use of designs, products, or technologies as alternatives to those prescribed, provided they result in substantially equivalent or greater *accessibility* and usability.

> **Advisory 103 Equivalent Facilitation.** The responsibility for demonstrating equivalent facilitation in the event of a challenge rests with the covered entity. With the exception of transit facilities, which are covered by regulations issued by the Department of Transportation, there is no process for certifying that an alternative design provides equivalent facilitation.

104 Conventions

104.1 Dimensions. Dimensions that are not stated as "maximum" or "minimum" are absolute.

104.1.1 Construction and Manufacturing Tolerances. All dimensions are subject to conventional industry tolerances except where the requirement is stated as a range with specific minimum and maximum end points.

> **Advisory 104.1.1 Construction and Manufacturing Tolerances.** Conventional industry tolerances recognized by this provision include those for field conditions and those that may be a necessary consequence of a particular manufacturing process. Recognized tolerances are not intended to apply to design work.
>
> It is good practice when specifying dimensions to avoid specifying a tolerance where dimensions are absolute. For example, if this document requires "1½ inches," avoid specifying "1½ inches plus or minus X inches."
>
> Where the requirement states a specified range, such as in Section 609.4 where grab bars must be installed between 33 inches and 36 inches above the floor, the range provides an adequate tolerance and therefore no tolerance outside of the range at either end point is permitted.
>
> Where a requirement is a minimum or a maximum dimension that does not have two specific minimum and maximum end points, tolerances may apply. Where an element is to be installed at the minimum or maximum permitted dimension, such as "15 inches minimum" or "5 pounds maximum", it would not be good practice to specify "5 pounds (plus X pounds) or 15 inches (minus X inches)." Rather, it would be good practice to specify a dimension less than the required maximum (or more than the required minimum) by the amount of the expected field or manufacturing tolerance and not to state any tolerance in conjunction with the specified dimension.
>
> Specifying dimensions in design in the manner described above will better ensure that facilities and elements accomplish the level of accessibility intended by these requirements. It will also more often produce an end result of strict and literal compliance with the stated requirements and eliminate enforcement difficulties and issues that might otherwise arise. Information on specific tolerances may be available from industry or trade organizations, code groups and building officials, and published references.

104.2 Calculation of Percentages. Where the required number of *elements* or *facilities* to be provided is determined by calculations of ratios or percentages and remainders or fractions result, the next greater whole number of such *elements* or *facilities* shall be provided. Where the determination of the required size or dimension of an *element* or *facility* involves ratios or percentages, rounding down for values less than one half shall be permitted.

104.3 Figures. Unless specifically stated otherwise, figures are provided for informational purposes only.

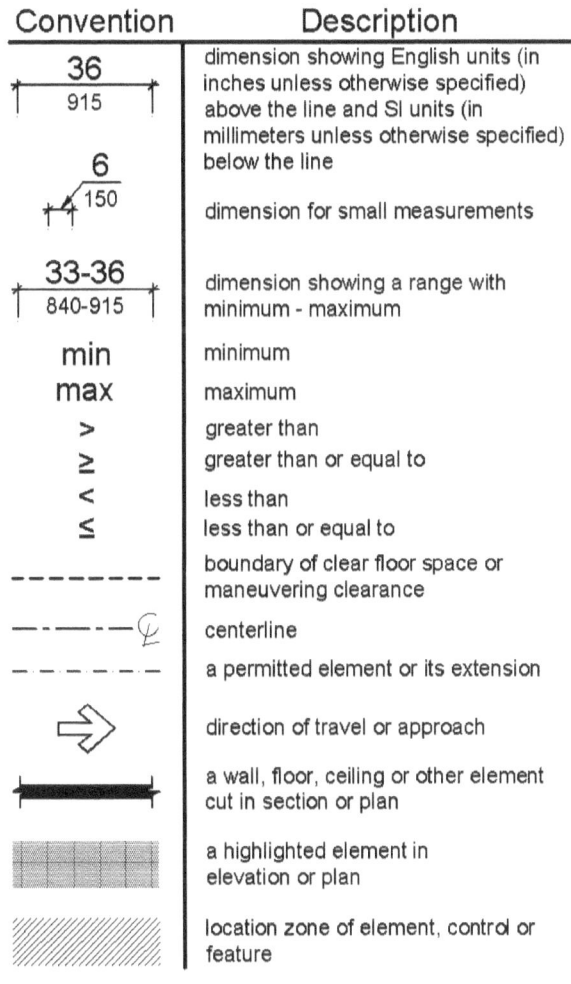

Figure 104
Graphic Convention for Figures

105 Referenced Standards

105.1 General. The standards listed in 105.2 are incorporated by reference in this document and are part of the requirements to the prescribed extent of each such reference. The Director of the Federal Register has approved these standards for incorporation by reference in accordance with 5 U.S.C. 552(a) and 1 CFR part 51. Copies of the referenced standards may be inspected at the Architectural and Transportation Barriers Compliance Board, 1331 F Street, NW, Suite 1000, Washington, DC 20004; at the Department of Justice, Civil Rights Division, Disability Rights Section, 1425 New York Avenue, NW, Washington, DC; at the Department of Transportation, 400 Seventh Street, SW, Room 10424, Washington DC; or at the National Archives and Records Administration (NARA). For information on the availability of this material at NARA, call (202) 741-6030, or go to http://www.archives.gov/federal_register/code_of_federal_regulations/ibr_locations.html.

105.2 Referenced Standards. The specific edition of the standards listed below are referenced in this document. Where differences occur between this document and the referenced standards, this document applies.

105.2.1 ANSI/BHMA. Copies of the referenced standards may be obtained from the Builders Hardware Manufacturers Association, 355 Lexington Avenue, 17th floor, New York, NY 10017 (http://www.buildershardware.com).

ANSI/BHMA A156.10-1999 American National Standard for Power Operated Pedestrian Doors (see 404.3).

ANSI/BHMA A156.19-1997 American National Standard for Power Assist and Low Energy Power Operated Doors (see 404.3, 408.3.2.1, and 409.3.1).

ANSI/BHMA A156.19-2002 American National Standard for Power Assist and Low Energy Power Operated Doors (see 404.3, 408.3.2.1, and 409.3.1).

> **Advisory 105.2.1 ANSI/BHMA.** ANSI/BHMA A156.10-1999 applies to power operated doors for pedestrian use which open automatically when approached by pedestrians. Included are provisions intended to reduce the chance of user injury or entrapment.
>
> ANSI/BHMA A156.19-1997 and A156.19-2002 applies to power assist doors, low energy power operated doors or low energy power open doors for pedestrian use not provided for in ANSI/BHMA A156.10 for Power Operated Pedestrian Doors. Included are provisions intended to reduce the chance of user injury or entrapment.

105.2.2 ASME. Copies of the referenced standards may be obtained from the American Society of Mechanical Engineers, Three Park Avenue, New York, New York 10016 (http://www.asme.org).

ASME A17.1- 2000 Safety Code for Elevators and Escalators, including ASME A17.1a-2002 Addenda and ASME A17.1b-2003 Addenda (see 407.1, 408.1, 409.1, and 810.9).

ASME A18.1-1999 Safety Standard for Platform Lifts and Stairway Chairlifts, including ASME A18.1a-2001 Addenda and ASME A18.1b-2001 Addenda (see 410.1).

ASME A18.1-2003 Safety Standard for Platform Lifts and Stairway Chairlifts, (see 410.1).

> **Advisory 105.2.2 ASME.** ASME A17.1-2000 is used by local jurisdictions throughout the United States for the design, construction, installation, operation, inspection, testing, maintenance, alteration, and repair of elevators and escalators. The majority of the requirements apply to the operational machinery not seen or used by elevator passengers. ASME A17.1 requires a two-way means of emergency communications in passenger elevators. This means of communication must connect with emergency or authorized personnel and not an automated answering system. The communication system must be push button activated. The activation button must be permanently identified with the word "HELP." A visual indication acknowledging the establishment of a communications link to authorized personnel must be provided. The visual indication must remain on until the call is terminated by authorized personnel. The building location, the elevator car number, and the need for assistance must be provided to authorized personnel answering the emergency call. The use of a handset by the communications system is prohibited. Only the authorized personnel answering the call can terminate the call. Operating instructions for the communications system must be provided in the elevator car.
>
> The provisions for escalators require that at least two flat steps be provided at the entrance and exit of every escalator and that steps on escalators be demarcated by yellow lines 2 inches wide maximum along the back and sides of steps.
>
> ASME A18.1-1999 and ASME A18.1-2003 address the design, construction, installation, operation, inspection, testing, maintenance and repair of lifts that are intended for transportation of persons with disabilities. Lifts are classified as: vertical platform lifts, inclined platform lifts, inclined stairway chairlifts, private residence vertical platform lifts, private residence inclined platform lifts, and private residence inclined stairway chairlifts.
>
> This document does not permit the use of inclined stairway chairlifts which do not provide platforms because such lifts require the user to transfer to a seat.
>
> ASME A18.1 contains requirements for runways, which are the spaces in which platforms or seats move. The standard includes additional provisions for runway enclosures, electrical equipment and wiring, structural support, headroom clearance (which is 80 inches minimum), lower level access ramps and pits. The enclosure walls not used for entry or exit are required to have a grab bar the full length of the wall on platform lifts. Access ramps are required to meet requirements similar to those for ramps in Chapter 4 of this document.
>
> Each of the lift types addressed in ASME A18.1 must meet requirements for capacity, load, speed, travel, operating devices, and control equipment. The maximum permitted height for operable parts is consistent with Section 308 of this document. The standard also addresses attendant operation. However, Section 410.1 of this document does not permit attendant operation.

105.2.3 ASTM. Copies of the referenced standards may be obtained from the American Society for Testing and Materials, 100 Bar Harbor Drive, West Conshohocken, Pennsylvania 19428 (http://www.astm.org).

ASTM F 1292-99 Standard Specification for Impact Attenuation of Surface Systems Under and Around Playground Equipment (see 1008.2.6.2).

ASTM F 1292-04 Standard Specification for Impact Attenuation of Surfacing Materials Within the Use Zone of Playground Equipment (see 1008.2.6.2).

ASTM F 1487-01 Standard Consumer Safety Performance Specification for Playground Equipment for Public Use (see 106.5).

ASTM F 1951-99 Standard Specification for Determination of Accessibility of Surface Systems Under and Around Playground Equipment (see 1008.2.6.1).

> **Advisory 105.2.3 ASTM.** ASTM F 1292-99 and ASTM F 1292-04 establish a uniform means to measure and compare characteristics of surfacing materials to determine whether materials provide a safe surface under and around playground equipment. These standards are referenced in the play areas requirements of this document when an accessible surface is required inside a play area use zone where a fall attenuating surface is also required. The standards cover the minimum impact attenuation requirements, when tested in accordance with Test Method F 355, for surface systems to be used under and around any piece of playground equipment from which a person may fall.
>
> ASTM F 1487-01 establishes a nationally recognized safety standard for public playground equipment to address injuries identified by the U.S. Consumer Product Safety Commission. It defines the use zone, which is the ground area beneath and immediately adjacent to a play structure or play equipment designed for unrestricted circulation around the equipment and on whose surface it is predicted that a user would land when falling from or exiting a play structure or equipment. The play areas requirements in this document reference the ASTM F 1487 standard when defining accessible routes that overlap use zones requiring fall attenuating surfaces. If the use zone of a playground is not entirely surfaced with an accessible material, at least one accessible route within the use zone must be provided from the perimeter to all accessible play structures or components within the playground.
>
> ASTM F 1951-99 establishes a uniform means to measure the characteristics of surface systems in order to provide performance specifications to select materials for use as an accessible surface under and around playground equipment. Surface materials that comply with this standard and are located in the use zone must also comply with ASTM F 1292. The test methods in this standard address access for children and adults who may traverse the surfacing to aid children who are playing. When a surface is tested it must have an average work per foot value for straight propulsion and for turning less than the average work per foot values for straight propulsion and for turning, respectively, on a hard, smooth surface with a grade of 7% (1:14).

105.2.4 ICC/IBC. Copies of the referenced standard may be obtained from the International Code Council, 5203 Leesburg Pike, Suite 600, Falls Church, Virginia 22041 (www.iccsafe.org).

International Building Code, 2000 Edition (see 207.1, 207.2, 216.4.2, 216.4.3, and 1005.2.1).

International Building Code, 2001 Supplement (see 207.1 and 207.2).

International Building Code, 2003 Edition (see 207.1, 207.2, 216.4.2, 216.4.3, and 1005.2.1).

> **Advisory 105.2.4 ICC/IBC.** International Building Code (IBC)-2000 (including 2001 Supplement to the International Codes) and IBC-2003 are referenced for means of egress, areas of refuge, and railings provided on fishing piers and platforms. At least one accessible means of egress is required for every accessible space and at least two accessible means of egress are required where more than one means of egress is required. The technical criteria for accessible means of egress allow the use of exit stairways and evacuation elevators when provided in conjunction with horizontal exits or areas of refuge. While typical elevators are not designed to be used during an emergency evacuation, evacuation elevators are designed with standby power and other features according to the elevator safety standard and can be used for the evacuation of individuals with disabilities. The IBC also provides requirements for areas of refuge, which are fire-rated spaces on levels above or below the exit discharge levels where people unable to use stairs can go to register a call for assistance and wait for evacuation.
>
> The recreation facilities requirements of this document references two sections in the IBC for fishing piers and platforms. An exception addresses the height of the railings, guards, or handrails where a fishing pier or platform is required to include a guard, railing, or handrail higher than 34 inches (865 mm) above the ground or deck surface.

105.2.5 NFPA. Copies of the referenced standards may be obtained from the National Fire Protection Association, 1 Batterymarch Park, Quincy, Massachusetts 02169-7471, (http://www.nfpa.org).

NFPA 72 National Fire Alarm Code, 1999 Edition (see 702.1 and 809.5.2).

NFPA 72 National Fire Alarm Code, 2002 Edition (see 702.1 and 809.5.2).

> **Advisory 105.2.5 NFPA.** NFPA 72-1999 and NFPA 72-2002 address the application, installation, performance, and maintenance of protective signaling systems and their components. The NFPA 72 incorporates Underwriters Laboratory (UL) 1971 by reference. The standard specifies the characteristics of audible alarms, such as placement and sound levels. However, Section 702 of these requirements limits the volume of an audible alarm to 110 dBA, rather than the maximum 120 dBA permitted by NFPA 72-1999.
>
> NFPA 72 specifies characteristics for visible alarms, such as flash frequency, color, intensity, placement, and synchronization. However, Section 702 of this document requires that visual alarm appliances be permanently installed. UL 1971 specifies intensity dispersion requirements for visible alarms. In particular, NFPA 72 requires visible alarms to have a light source that is clear or white and has polar dispersion complying with UL 1971.

106 Definitions

106.1 General. For the purpose of this document, the terms defined in 106.5 have the indicated meaning.

> **Advisory 106.1 General.** Terms defined in Section 106.5 are italicized in the text of this document.

106.2 Terms Defined in Referenced Standards. Terms not defined in 106.5 or in regulations issued by the Department of Justice and the Department of Transportation to implement the Americans with Disabilities Act, but specifically defined in a referenced standard, shall have the specified meaning from the referenced standard unless otherwise stated.

106.3 Undefined Terms. The meaning of terms not specifically defined in 106.5 or in regulations issued by the Department of Justice and the Department of Transportation to implement the Americans with Disabilities Act or in referenced standards shall be as defined by collegiate dictionaries in the sense that the context implies.

106.4 Interchangeability. Words, terms and phrases used in the singular include the plural and those used in the plural include the singular.

106.5 Defined Terms.

Accessible. A *site*, *building*, *facility*, or portion thereof that complies with this part.

Accessible Means of Egress. A continuous and unobstructed way of egress travel from any point in a *building* or *facility* that provides an *accessible* route to an area of refuge, a horizontal exit, or a *public way*.

Addition. An expansion, extension, or increase in the gross floor area or height of a *building* or *facility*.

Administrative Authority. A governmental agency that adopts or enforces regulations and guidelines for the design, construction, or *alteration* of *buildings* and *facilities*.

Alteration. A change to a *building* or *facility* that affects or could affect the usability of the *building* or *facility* or portion thereof. *Alterations* include, but are not limited to, remodeling, renovation, rehabilitation, reconstruction, historic restoration, resurfacing of *circulation paths* or *vehicular ways*, changes or rearrangement of the structural parts or *elements*, and changes or rearrangement in the plan configuration of walls and full-height partitions. Normal maintenance, reroofing, painting or wallpapering, or changes to mechanical and electrical systems are not *alterations* unless they affect the usability of the *building* or *facility*.

Amusement Attraction. Any *facility*, or portion of a *facility*, located within an amusement park or theme park which provides amusement without the use of an amusement device. *Amusement attractions* include, but are not limited to, fun houses, barrels, and other attractions without seats.

Amusement Ride. A system that moves persons through a fixed course within a defined area for the purpose of amusement.

Amusement Ride Seat. A seat that is built-in or mechanically fastened to an *amusement ride* intended to be occupied by one or more passengers.

Area of Sport Activity. That portion of a room or *space* where the play or practice of a sport occurs.

Assembly Area. A *building* or *facility*, or portion thereof, used for the purpose of entertainment, educational or civic gatherings, or similar purposes. For the purposes of these requirements, *assembly areas* include, but are not limited to, classrooms, lecture halls, courtrooms, public meeting rooms, public hearing rooms, legislative chambers, motion picture houses, auditoria, theaters, playhouses, dinner theaters, concert halls, centers for the performing arts, amphitheaters, arenas, stadiums, grandstands, or convention centers.

Assistive Listening System (ALS). An amplification system utilizing transmitters, receivers, and coupling devices to bypass the acoustical *space* between a sound source and a listener by means of induction loop, radio frequency, infrared, or direct-wired equipment.

Boarding Pier. A portion of a pier where a boat is temporarily secured for the purpose of embarking or disembarking.

Boat Launch Ramp. A sloped surface designed for launching and retrieving trailered boats and other water craft to and from a body of water.

Boat Slip. That portion of a pier, main pier, finger pier, or float where a boat is moored for the purpose of berthing, embarking, or disembarking.

Building. Any structure used or intended for supporting or sheltering any use or occupancy.

Catch Pool. A pool or designated section of a pool used as a terminus for water slide flumes.

Characters. Letters, numbers, punctuation marks and typographic symbols.

Children's Use. Describes *spaces* and *elements* specifically designed for use primarily by people 12 years old and younger.

Circulation Path. An exterior or interior way of passage provided for pedestrian travel, including but not limited to, *walks*, hallways, courtyards, elevators, platform lifts, *ramps*, stairways, and landings.

Closed-Circuit Telephone. A telephone with a dedicated line such as a house phone, courtesy phone or phone that must be used to gain entry to a *facility*.

Common Use. Interior or exterior *circulation paths*, rooms, *spaces*, or *elements* that are not for *public use* and are made available for the shared use of two or more people.

Cross Slope. The slope that is perpendicular to the direction of travel (see *running slope*).

Curb Ramp. A short *ramp* cutting through a curb or built up to it.

Detectable Warning. A standardized surface feature built in or applied to walking surfaces or other *elements* to warn of hazards on a *circulation path*.

Element. An architectural or mechanical component of a *building*, *facility*, *space*, or *site*.

Elevated Play Component. A *play component* that is approached above or below grade and that is part of a composite play structure consisting of two or more *play components* attached or functionally linked to create an integrated unit providing more than one play activity.

Employee Work Area. All or any portion of a *space* used only by employees and used only for work. Corridors, toilet rooms, kitchenettes and break rooms are not *employee work areas*.

Entrance. Any access point to a *building* or portion of a *building* or *facility* used for the purpose of entering. An *entrance* includes the approach *walk*, the vertical access leading to the *entrance* platform, the *entrance* platform itself, vestibule if provided, the entry door or gate, and the hardware of the entry door or gate.

Facility. All or any portion of *buildings*, structures, *site* improvements, *elements*, and pedestrian routes or *vehicular ways* located on a *site*.

Gangway. A variable-sloped pedestrian walkway that links a fixed structure or land with a floating structure. *Gangways* that connect to vessels are not addressed by this document.

Golf Car Passage. A continuous passage on which a motorized golf car can operate.

Ground Level Play Component. A *play component* that is approached and exited at the ground level.

Key Station. Rapid and light rail stations, and commuter rail stations, as defined under criteria established by the Department of Transportation in 49 CFR 37.47 and 49 CFR 37.51, respectively.

Mail Boxes. Receptacles for the receipt of documents, packages, or other deliverable matter. *Mail boxes* include, but are not limited to, post office boxes and receptacles provided by commercial mail-receiving agencies, apartment *facilities*, or schools.

Marked Crossing. A crosswalk or other identified path intended for pedestrian use in crossing a *vehicular way*.

Mezzanine. An intermediate level or levels between the floor and ceiling of any *story* with an aggregate floor area of not more than one-third of the area of the room or *space* in which the level or levels are located. *Mezzanines* have sufficient elevation that *space* for human occupancy can be provided on the floor below.

Occupant Load. The number of persons for which the means of egress of a *building* or portion of a *building* is designed.

Operable Part. A component of an *element* used to insert or withdraw objects, or to activate, deactivate, or adjust the *element*.

Pictogram. A pictorial symbol that represents activities, *facilities*, or concepts.

Play Area. A portion of a *site* containing *play components* designed and constructed for children.

Play Component. An *element* intended to generate specific opportunities for play, socialization, or learning. *Play components* are manufactured or natural; and are stand-alone or part of a composite play structure.

Private Building or Facility. A place of public accommodation or a commercial *building* or *facility* subject to title III of the ADA and 28 CFR part 36 or a transportation *building* or *facility* subject to title III of the ADA and 49 CFR 37.45.

Public Building or Facility. A *building* or *facility* or portion of a *building* or *facility* designed, constructed, or *altered* by, on behalf of, or for the use of a public entity subject to title II of the ADA and 28 CFR part 35 or to title II of the ADA and 49 CFR 37.41 or 37.43.

Public Entrance. An *entrance* that is not a *service entrance* or a *restricted entrance*.

Public Use. Interior or exterior rooms, *spaces*, or *elements* that are made available to the public. *Public use* may be provided at a *building* or *facility* that is privately or publicly owned.

Public Way. Any street, alley or other parcel of land open to the outside air leading to a public street, which has been deeded, dedicated or otherwise permanently appropriated to the public for *public use* and which has a clear width and height of not less than 10 feet (3050 mm).

Qualified Historic Building or Facility. A *building* or *facility* that is listed in or eligible for listing in the National Register of Historic Places, or designated as historic under an appropriate State or local law.

Ramp. A walking surface that has a *running slope* steeper than 1:20.

Residential Dwelling Unit. A unit intended to be used as a residence, that is primarily long-term in nature. *Residential dwelling units* do not include *transient lodging*, inpatient medical care, licensed long-term care, and detention or correctional *facilities*.

Restricted Entrance. An *entrance* that is made available for *common use* on a controlled basis but not *public use* and that is not a *service entrance*.

Running Slope. The slope that is parallel to the direction of travel (see *cross slope*).

Self-Service Storage. *Building* or *facility* designed and used for the purpose of renting or leasing individual storage *spaces* to customers for the purpose of storing and removing personal property on a self-service basis.

Service Entrance. An *entrance* intended primarily for delivery of goods or services.

Site. A parcel of land bounded by a property line or a designated portion of a public right-of-way.

Soft Contained Play Structure. A play structure made up of one or more *play components* where the user enters a fully enclosed play environment that utilizes pliable materials, such as plastic, netting, or fabric.

Space. A definable area, such as a room, toilet room, hall, *assembly area*, *entrance*, storage room, alcove, courtyard, or lobby.

Story. That portion of a *building* or *facility* designed for human occupancy included between the upper surface of a floor and upper surface of the floor or roof next above. A *story* containing one or more *mezzanines* has more than one floor level.

Structural Frame. The columns and the girders, beams, and trusses having direct connections to the columns and all other members that are essential to the stability of the *building* or *facility* as a whole.

Tactile. An object that can be perceived using the sense of touch.

Technically Infeasible. With respect to an *alteration* of a *building* or a *facility*, something that has little likelihood of being accomplished because existing structural conditions would require removing or *altering* a load-bearing member that is an essential part of the *structural frame*; or because other existing physical or *site* constraints prohibit modification or *addition* of *elements*, *spaces*, or features that are in full and strict compliance with the minimum requirements.

Teeing Ground. In golf, the starting place for the hole to be played.

Transfer Device. Equipment designed to facilitate the transfer of a person from a wheelchair or other mobility aid to and from an *amusement ride seat*.

Transient Lodging. A *building* or *facility* containing one or more guest room(s) for sleeping that provides accommodations that are primarily short-term in nature. *Transient lodging* does not include *residential dwelling units* intended to be used as a residence, inpatient medical care *facilities*, licensed long-term care *facilities*, detention or correctional *facilities*, or *private buildings or facilities* that contain not more than five rooms for rent or hire and that are actually occupied by the proprietor as the residence of such proprietor.

Transition Plate. A sloping pedestrian walking surface located at the end(s) of a *gangway*.

TTY. An abbreviation for teletypewriter. Machinery that employs interactive text-based communication through the transmission of coded signals across the telephone network. *TTYs* may include, for example, devices known as TDDs (telecommunication display devices or telecommunication devices for deaf persons) or computers with special modems. *TTYs* are also called text telephones.

Use Zone. The ground level area beneath and immediately adjacent to a play structure or play equipment that is designated by ASTM F 1487 (incorporated by reference, see "Referenced Standards" in Chapter 1) for unrestricted circulation around the play equipment and where it is predicted that a user would land when falling from or exiting the play equipment.

Vehicular Way. A route provided for vehicular traffic, such as in a street, driveway, or parking *facility*.

Walk. An exterior prepared surface for pedestrian use, including pedestrian areas such as plazas and courts.

Wheelchair Space. *Space* for a single wheelchair and its occupant.

Work Area Equipment. Any machine, instrument, engine, motor, pump, conveyor, or other apparatus used to perform work. As used in this document, this term shall apply only to equipment that is permanently installed or built-in in *employee work areas*. *Work area equipment* does not include passenger elevators and other accessible means of vertical transportation.

ADA CHAPTER 2: SCOPING REQUIREMENTS

201 Application

201.1 Scope. All areas of newly designed and newly constructed *buildings* and *facilities* and *altered* portions of existing *buildings* and *facilities* shall comply with these requirements.

> **Advisory 201.1 Scope.** These requirements are to be applied to all areas of a facility unless exempted, or where scoping limits the number of multiple elements required to be accessible. For example, not all medical care patient rooms are required to be accessible; those that are not required to be accessible are not required to comply with these requirements. However, common use and public use spaces such as recovery rooms, examination rooms, and cafeterias are not exempt from these requirements and must be accessible.

201.2 Application Based on Building or Facility Use. Where a *site*, *building*, *facility*, room, or *space* contains more than one use, each portion shall comply with the applicable requirements for that use.

201.3 Temporary and Permanent Structures. These requirements shall apply to temporary and permanent *buildings* and *facilities*.

> **Advisory 201.3 Temporary and Permanent Structures.** Temporary buildings or facilities covered by these requirements include, but are not limited to, reviewing stands, temporary classrooms, bleacher areas, stages, platforms and daises, fixed furniture systems, wall systems, and exhibit areas, temporary banking facilities, and temporary health screening facilities. Structures and equipment directly associated with the actual processes of construction are not required to be accessible as permitted in 203.2.

202 Existing Buildings and Facilities

202.1 General. *Additions* and *alterations* to existing *buildings* or *facilities* shall comply with 202.

202.2 Additions. Each *addition* to an existing *building* or *facility* shall comply with the requirements for new construction. Each *addition* that affects or could affect the usability of or access to an area containing a primary function shall comply with 202.4.

202.3 Alterations. Where existing *elements* or *spaces* are *altered*, each *altered element* or *space* shall comply with the applicable requirements of Chapter 2.
    EXCEPTIONS: 1. Unless required by 202.4, where *elements* or *spaces* are *altered* and the *circulation path* to the *altered element* or *space* is not *altered*, an *accessible* route shall not be required.
    2. In *alterations*, where compliance with applicable requirements is *technically infeasible*, the *alteration* shall comply with the requirements to the maximum extent feasible.

3. *Residential dwelling units* not required to be *accessible* in compliance with a standard issued pursuant to the Americans with Disabilities Act or Section 504 of the Rehabilitation Act of 1973, as amended, shall not be required to comply with 202.3.

> **Advisory 202.3 Alterations.** Although covered entities are permitted to limit the scope of an alteration to individual elements, the alteration of multiple elements within a room or space may provide a cost-effective opportunity to make the entire room or space accessible. Any elements or spaces of the building or facility that are required to comply with these requirements must be made accessible within the scope of the alteration, to the maximum extent feasible. If providing accessibility in compliance with these requirements for people with one type of disability (e.g., people who use wheelchairs) is not feasible, accessibility must still be provided in compliance with the requirements for people with other types of disabilities (e.g., people who have hearing impairments or who have vision impairments) to the extent that such accessibility is feasible.

202.3.1 Prohibited Reduction in Access. An *alteration* that decreases or has the effect of decreasing the *accessibility* of a *building* or *facility* below the requirements for new construction at the time of the *alteration* is prohibited.

202.3.2 Extent of Application. An *alteration* of an existing *element*, *space*, or area of a *building* or *facility* shall not impose a requirement for *accessibility* greater than required for new construction.

202.4 Alterations Affecting Primary Function Areas. In addition to the requirements of 202.3, an *alteration* that affects or could affect the usability of or access to an area containing a primary function shall be made so as to ensure that, to the maximum extent feasible, the path of travel to the *altered* area, including the rest rooms, telephones, and drinking fountains serving the *altered* area, are readily *accessible* to and usable by individuals with disabilities, unless such *alteration*s are disproportionate to the overall *alterations* in terms of cost and scope as determined under criteria established by the Attorney General. In existing transportation *facilities*, an area of primary function shall be as defined under regulations published by the Secretary of the Department of Transportation or the Attorney General.

> EXCEPTION: *Residential dwelling units* shall not be required to comply with 202.4.

> **Advisory 202.4 Alterations Affecting Primary Function Areas.** An area of a building or facility containing a major activity for which the building or facility is intended is a primary function area. Department of Justice ADA regulations state, "Alterations made to provide an accessible path of travel to the altered area will be deemed disproportionate to the overall alteration when the cost exceeds 20% of the cost of the alteration to the primary function area." (28 CFR 36.403 (f)(1)). See also Department of Transportation ADA regulations, which use similar concepts in the context of public sector transportation facilities (49 CFR 37.43 (e)(1)).
>
> There can be multiple areas containing a primary function in a single building. Primary function areas are not limited to public use areas. For example, both a bank lobby and the bank's employee areas such as the teller areas and walk-in safe are primary function areas.

> **Advisory 202.4 Alterations Affecting Primary Function Areas (Continued).** Also, mixed use facilities may include numerous primary function areas for each use. Areas containing a primary function do not include: mechanical rooms, boiler rooms, supply storage rooms, employee lounges or locker rooms, janitorial closets, entrances, corridors, or restrooms.

202.5 Alterations to Qualified Historic Buildings and Facilities. *Alterations* to a *qualified historic building or facility* shall comply with 202.3 and 202.4.

    EXCEPTION: Where the State Historic Preservation Officer or Advisory Council on Historic Preservation determines that compliance with the requirements for *accessible* routes, *entrances*, or toilet *facilities* would threaten or destroy the historic significance of the *building* or *facility*, the exceptions for *alterations* to *qualified historic buildings or facilities* for that *element* shall be permitted to apply.

> **Advisory 202.5 Alterations to Qualified Historic Buildings and Facilities Exception.** State Historic Preservation Officers are State appointed officials who carry out certain responsibilities under the National Historic Preservation Act. State Historic Preservation Officers consult with Federal and State agencies, local governments, and private entities on providing access and protecting significant elements of qualified historic buildings and facilities. There are exceptions for alterations to qualified historic buildings and facilities for accessible routes (206.2.1 Exception 1 and 206.2.3 Exception 7); entrances (206.4 Exception 2); and toilet facilities (213.2 Exception 2). When an entity believes that compliance with the requirements for any of these elements would threaten or destroy the historic significance of the building or facility, the entity should consult with the State Historic Preservation Officer. If the State Historic Preservation Officer agrees that compliance with the requirements for a specific element would threaten or destroy the historic significance of the building or facility, use of the exception is permitted. Public entities have an additional obligation to achieve program accessibility under the Department of Justice ADA regulations. See 28 CFR 35.150. These regulations require public entities that operate historic preservation programs to give priority to methods that provide physical access to individuals with disabilities. If alterations to a qualified historic building or facility to achieve program accessibility would threaten or destroy the historic significance of the building or facility, fundamentally alter the program, or result in undue financial or administrative burdens, the Department of Justice ADA regulations allow alternative methods to be used to achieve program accessibility. In the case of historic preservation programs, such as an historic house museum, alternative methods include using audio-visual materials to depict portions of the house that cannot otherwise be made accessible. In the case of other qualified historic properties, such as an historic government office building, alternative methods include relocating programs and services to accessible locations. The Department of Justice ADA regulations also allow public entities to use alternative methods when altering qualified historic buildings or facilities in the rare situations where the State Historic Preservation Officer determines that it is not feasible to provide physical access using the exceptions permitted in Section 202.5 without threatening or destroying the historic significance of the building or facility. See 28 CFR 35.151(d).

> **Advisory 202.5 Alterations to Qualified Historic Buildings and Facilities Exception (Continued).** The AccessAbility Office at the National Endowment for the Arts (NEA) provides a variety of resources for museum operators and historic properties including: the Design for Accessibility Guide and the Disability Symbols. Contact NEA about these and other resources at (202) 682-5532 or www.arts.gov.

203 General Exceptions

203.1 General. *Sites, buildings, facilities,* and *elements* are exempt from these requirements to the extent specified by 203.

203.2 Construction Sites. Structures and *sites* directly associated with the actual processes of construction, including but not limited to, scaffolding, bridging, materials hoists, materials storage, and construction trailers shall not be required to comply with these requirements or to be on an *accessible* route. Portable toilet units provided for use exclusively by construction personnel on a construction *site* shall not be required to comply with 213 or to be on an *accessible* route.

203.3 Raised Areas. Areas raised primarily for purposes of security, life safety, or fire safety, including but not limited to, observation or lookout galleries, prison guard towers, fire towers, or life guard stands shall not be required to comply with these requirements or to be on an *accessible* route.

203.4 Limited Access Spaces. *Spaces* accessed only by ladders, catwalks, crawl *spaces,* or very narrow passageways shall not be required to comply with these requirements or to be on an *accessible* route.

203.5 Machinery Spaces. *Spaces* frequented only by service personnel for maintenance, repair, or occasional monitoring of equipment shall not be required to comply with these requirements or to be on an *accessible* route. Machinery *spaces* include, but are not limited to, elevator pits or elevator penthouses; mechanical, electrical or communications equipment rooms; piping or equipment catwalks; water or sewage treatment pump rooms and stations; electric substations and transformer vaults; and highway and tunnel utility *facilities*.

203.6 Single Occupant Structures. Single occupant structures accessed only by passageways below grade or elevated above standard curb height, including but not limited to, toll booths that are accessed only by underground tunnels, shall not be required to comply with these requirements or to be on an *accessible* route.

203.7 Detention and Correctional Facilities. In detention and correctional *facilities, common use* areas that are used only by inmates or detainees and security personnel and that do not serve holding cells or housing cells required to comply with 232, shall not be required to comply with these requirements or to be on an *accessible* route.

203.8 Residential Facilities. In residential *facilities, common use* areas that do not serve *residential dwelling units* required to provide mobility features complying with 809.2 through 809.4 shall not be required to comply with these requirements or to be on an *accessible* route.

203.9 Employee Work Areas. *Spaces* and *elements* within *employee work areas* shall only be required to comply with 206.2.8, 207.1, and 215.3 and shall be designed and constructed so that individuals with disabilities can approach, enter, and exit the *employee work area*. *Employee work areas*, or portions of *employee work areas*, other than raised courtroom stations, that are less than 300 square feet (28 m^2) and elevated 7 inches (180 mm) or more above the finish floor or ground where the elevation is essential to the function of the *space* shall not be required to comply with these requirements or to be on an *accessible* route.

> **Advisory 203.9 Employee Work Areas.** Although areas used exclusively by employees for work are not required to be fully accessible, consider designing such areas to include non-required turning spaces, and provide accessible elements whenever possible. Under the ADA, employees with disabilities are entitled to reasonable accommodations in the workplace; accommodations can include alterations to spaces within the facility. Designing employee work areas to be more accessible at the outset will avoid more costly retrofits when current employees become temporarily or permanently disabled, or when new employees with disabilities are hired. Contact the Equal Employment Opportunity Commission (EEOC) at www.eeoc.gov for information about title I of the ADA prohibiting discrimination against people with disabilities in the workplace.

203.10 Raised Refereeing, Judging, and Scoring Areas. Raised structures used solely for refereeing, judging, or scoring a sport shall not be required to comply with these requirements or to be on an *accessible* route.

203.11 Water Slides. Water slides shall not be required to comply with these requirements or to be on an *accessible* route.

203.12 Animal Containment Areas. Animal containment areas that are not for *public use* shall not be required to comply with these requirements or to be on an *accessible* route.

> **Advisory 203.12 Animal Containment Areas.** Public circulation routes where animals may travel, such as in petting zoos and passageways alongside animal pens in State fairs, are not eligible for the exception.

203.13 Raised Boxing or Wrestling Rings. Raised boxing or wrestling rings shall not be required to comply with these requirements or to be on an *accessible* route.

203.14 Raised Diving Boards and Diving Platforms. Raised diving boards and diving platforms shall not be required to comply with these requirements or to be on an *accessible* route.

204 Protruding Objects

204.1 General. Protruding objects on *circulation paths* shall comply with 307.
    EXCEPTIONS: 1. Within *areas of sport activity*, protruding objects on *circulation paths* shall not be required to comply with 307.
    2. Within *play areas*, protruding objects on *circulation paths* shall not be required to comply with 307 provided that ground level *accessible* routes provide vertical clearance in compliance with 1008.2.

205 Operable Parts

205.1 General. *Operable parts* on *accessible elements*, *accessible* routes, and in *accessible* rooms and *spaces* shall comply with 309.

> **EXCEPTIONS: 1.** *Operable parts* that are intended for use only by service or maintenance personnel shall not be required to comply with 309.
> **2.** Electrical or communication receptacles serving a dedicated use shall not be required to comply with 309.
> **3.** Where two or more outlets are provided in a kitchen above a length of counter top that is uninterrupted by a sink or appliance, one outlet shall not be required to comply with 309.
> **4.** Floor electrical receptacles shall not be required to comply with 309.
> **5.** HVAC diffusers shall not be required to comply with 309.
> **6.** Except for light switches, where redundant controls are provided for a single *element*, one control in each *space* shall not be required to comply with 309.
> **7.** Cleats and other boat securement devices shall not be required to comply with 309.3.
> **8.** Exercise machines and exercise equipment shall not be required to comply with 309.

> **Advisory 205.1 General.** Controls covered by 205.1 include, but are not limited to, light switches, circuit breakers, duplexes and other convenience receptacles, environmental and appliance controls, plumbing fixture controls, and security and intercom systems.

206 Accessible Routes

206.1 General. *Accessible* routes shall be provided in accordance with 206 and shall comply with Chapter 4.

206.2 Where Required. *Accessible* routes shall be provided where required by 206.2.

206.2.1 Site Arrival Points. At least one *accessible* route shall be provided within the *site* from *accessible* parking *spaces* and *accessible* passenger loading zones; public streets and sidewalks; and public transportation stops to the *accessible building* or *facility entrance* they serve.

> **EXCEPTIONS: 1.** Where exceptions for *alterations* to *qualified historic buildings or facilities* are permitted by 202.5, no more than one *accessible* route from a *site* arrival point to an *accessible entrance* shall be required.
> **2.** An *accessible* route shall not be required between *site* arrival points and the *building* or *facility entrance* if the only means of access between them is a *vehicular way* not providing pedestrian access.

> **Advisory 206.2.1 Site Arrival Points.** Each site arrival point must be connected by an accessible route to the accessible building entrance or entrances served. Where two or more similar site arrival points, such as bus stops, serve the same accessible entrance or entrances, both bus stops must be on accessible routes. In addition, the accessible routes must serve all of the accessible entrances on the site.

> **Advisory 206.2.1 Site Arrival Points Exception 2.** Access from site arrival points may include vehicular ways. Where a vehicular way, or a portion of a vehicular way, is provided for pedestrian travel, such as within a shopping center or shopping mall parking lot, this exception does not apply.

206.2.2 Within a Site. At least one *accessible* route shall connect *accessible buildings, accessible facilities, accessible elements,* and *accessible spaces* that are on the same *site*.
EXCEPTION: An *accessible* route shall not be required between *accessible buildings, accessible facilities, accessible elements,* and *accessible spaces* if the only means of access between them is a *vehicular way* not providing pedestrian access.

> **Advisory 206.2.2 Within a Site.** An accessible route is required to connect to the boundary of each area of sport activity. Examples of areas of sport activity include: soccer fields, basketball courts, baseball fields, running tracks, skating rinks, and the area surrounding a piece of gymnastic equipment. While the size of an area of sport activity may vary from sport to sport, each includes only the space needed to play. Where multiple sports fields or courts are provided, an accessible route is required to each field or area of sport activity.

206.2.3 Multi-Story Buildings and Facilities. At least one *accessible* route shall connect each *story* and *mezzanine* in multi-*story buildings* and *facilities*.
EXCEPTIONS: 1. In *private buildings or facilities* that are less than three *stories* or that have less than 3000 square feet (279 m^2) per *story*, an *accessible* route shall not be required to connect *stories* provided that the *building* or *facility* is not a shopping center, a shopping mall, the professional office of a health care provider, a terminal, depot or other station used for specified public transportation, an airport passenger terminal, or another type of *facility* as determined by the Attorney General.
2. Where a two *story public building or facility* has one *story* with an *occupant load* of five or fewer persons that does not contain *public use space*, that *story* shall not be required to be connected to the *story* above or below.
3. In detention and correctional *facilities*, an *accessible* route shall not be required to connect *stories* where cells with mobility features required to comply with 807.2, all *common use* areas serving cells with mobility features required to comply with 807.2, and all *public use* areas are on an *accessible* route.
4. In residential *facilities*, an *accessible* route shall not be required to connect *stories* where *residential dwelling units* with mobility features required to comply with 809.2 through 809.4, all *common use* areas serving *residential dwelling units* with mobility features required to comply with 809.2 through 809.4, and *public use* areas serving *residential dwelling units* are on an *accessible* route.
5. Within multi-*story transient lodging* guest rooms with mobility features required to comply with 806.2, an *accessible* route shall not be required to connect *stories* provided that *spaces* complying with 806.2 are on an *accessible* route and sleeping accommodations for two persons minimum are provided on a *story* served by an accessible route.
6. In air traffic control towers, an *accessible* route shall not be required to serve the cab and the floor immediately below the cab.

7. Where exceptions for *alterations* to *qualified historic buildings or facilities* are permitted by 202.5, an *accessible* route shall not be required to *stories* located above or below the *accessible story*.

> **Advisory 206.2.3 Multi-Story Buildings and Facilities.** Spaces and elements located on a level not required to be served by an accessible route must fully comply with this document. While a mezzanine may be a change in level, it is not a story. If an accessible route is required to connect stories within a building or facility, the accessible route must serve all mezzanines.
>
> **Advisory 206.2.3 Multi-Story Buildings and Facilities Exception 4.** Where common use areas are provided for the use of residents, it is presumed that all such common use areas "serve" accessible dwelling units unless use is restricted to residents occupying certain dwelling units. For example, if all residents are permitted to use all laundry rooms, then all laundry rooms "serve" accessible dwelling units. However, if the laundry room on the first floor is restricted to use by residents on the first floor, and the second floor laundry room is for use by occupants of the second floor, then first floor accessible units are "served" only by laundry rooms on the first floor. In this example, an accessible route is not required to the second floor provided that all accessible units and all common use areas serving them are on the first floor.

206.2.3.1 Stairs and Escalators in Existing Buildings. In *alterations* and *additions*, where an escalator or stair is provided where none existed previously and major structural modifications are necessary for the installation, an *accessible* route shall be provided between the levels served by the escalator or stair unless exempted by 206.2.3 Exceptions 1 through 7.

206.2.4 Spaces and Elements. At least one *accessible* route shall connect *accessible building* or *facility entrances* with all *accessible spaces* and *elements* within the *building* or *facility* which are otherwise connected by a *circulation path* unless exempted by 206.2.3 Exceptions 1 through 7.
  EXCEPTIONS: 1. Raised courtroom stations, including judges' benches, clerks' stations, bailiffs' stations, deputy clerks' stations, and court reporters' stations shall not be required to provide vertical access provided that the required clear floor *space*, maneuvering *space*, and, if appropriate, electrical service are installed at the time of initial construction to allow future installation of a means of vertical access complying with 405, 407, 408, or 410 without requiring substantial reconstruction of the *space*.
  2. In *assembly areas* with fixed seating required to comply with 221, an *accessible* route shall not be required to serve fixed seating where *wheelchair spaces* required to be on an *accessible* route are not provided.
  3. *Accessible* routes shall not be required to connect *mezzanines* where *buildings* or *facilities* have no more than one story. In addition, *accessible* routes shall not be required to connect stories or *mezzanines* where multi-story *buildings* or *facilities* are exempted by 206.2.3 Exceptions 1 through 7.

> **Advisory 206.2.4 Spaces and Elements.** Accessible routes must connect all spaces and elements required to be accessible including, but not limited to, raised areas and speaker platforms.
>
> **Advisory 206.2.4 Spaces and Elements Exception 1.** The exception does not apply to areas that are likely to be used by members of the public who are not employees of the court such as jury areas, attorney areas, or witness stands.

206.2.5 Restaurants and Cafeterias. In restaurants and cafeterias, an *accessible* route shall be provided to all dining areas, including raised or sunken dining areas, and outdoor dining areas.
 EXCEPTIONS: 1. In *buildings or facilities* not required to provide an *accessible* route between *stories*, an *accessible* route shall not be required to a *mezzanine* dining area where the *mezzanine* contains less than 25 percent of the total combined area for seating and dining and where the same decor and services are provided in the *accessible* area.
 2. In *alterations*, an *accessible* route shall not be required to existing raised or sunken dining areas, or to all parts of existing outdoor dining areas where the same services and decor are provided in an *accessible space* usable by the public and not restricted to use by people with disabilities.
 3. In sports *facilities*, tiered dining areas providing seating required to comply with 221 shall be required to have *accessible* routes serving at least 25 percent of the dining area provided that *accessible* routes serve seating complying with 221 and each tier is provided with the same services.

> **Advisory 206.2.5 Restaurants and Cafeterias Exception 2.** Examples of "same services" include, but are not limited to, bar service, rooms having smoking and non-smoking sections, lotto and other table games, carry-out, and buffet service. Examples of "same decor" include, but are not limited to, seating at or near windows and railings with views, areas designed with a certain theme, party and banquet rooms, and rooms where entertainment is provided.

206.2.6 Performance Areas. Where a *circulation path* directly connects a performance area to an assembly seating area, an *accessible* route shall directly connect the assembly seating area with the performance area. An *accessible* route shall be provided from performance areas to ancillary areas or *facilities* used by performers unless exempted by 206.2.3 Exceptions 1 through 7.

206.2.7 Press Boxes. Press boxes in *assembly areas* shall be on an *accessible* route.
 EXCEPTIONS: 1. An *accessible* route shall not be required to press boxes in bleachers that have points of entry at only one level provided that the aggregate area of all press boxes is 500 square feet (46 m^2) maximum.
 2. An *accessible* route shall not be required to free-standing press boxes that are elevated above grade 12 feet (3660 mm) minimum provided that the aggregate area of all press boxes is 500 square feet (46 m^2) maximum.

> **Advisory 206.2.7 Press Boxes Exception 2.** Where a facility contains multiple assembly areas, the aggregate area of the press boxes in each assembly area is to be calculated separately. For example, if a university has a soccer stadium with three press boxes elevated 12 feet (3660 mm) or more above grade and each press box is 150 square feet (14 m^2), then the aggregate area of the soccer stadium press boxes is less than 500 square feet (46 m^2) and Exception 2 applies to the soccer stadium. If that same university also has a football stadium with two press boxes elevated 12 feet (3660 mm) or more above grade and one press box is 250 square feet (23 m^2), and the second is 275 square feet (26 m^2), then the aggregate area of the football stadium press boxes is more than 500 square feet (46 m^2) and Exception 2 does not apply to the football stadium.

206.2.8 Employee Work Areas. *Common use circulation paths* within *employee work areas* shall comply with 402.

EXCEPTIONS: 1. *Common use circulation paths* located within *employee work areas* that are less than 1000 square feet (93 m^2) and defined by permanently installed partitions, counters, casework, or furnishings shall not be required to comply with 402.

2. *Common use circulation paths* located within *employee work areas* that are an integral component of *work area equipment* shall not be required to comply with 402.

3. *Common use circulation paths* located within exterior *employee work areas* that are fully exposed to the weather shall not be required to comply with 402.

> **Advisory 206.2.8 Employee Work Areas Exception 1.** Modular furniture that is not permanently installed is not directly subject to these requirements. The Department of Justice ADA regulations provide additional guidance regarding the relationship between these requirements and elements that are not part of the built environment. Additionally, the Equal Employment Opportunity Commission (EEOC) implements title I of the ADA which requires non-discrimination in the workplace. EEOC can provide guidance regarding employers' obligations to provide reasonable accommodations for employees with disabilities.
>
> **Advisory 206.2.8 Employee Work Areas Exception 2.** Large pieces of equipment, such as electric turbines or water pumping apparatus, may have stairs and elevated walkways used for overseeing or monitoring purposes which are physically part of the turbine or pump. However, passenger elevators used for vertical transportation between stories are not considered "work area equipment" as defined in Section 106.5.

206.2.9 Amusement Rides. *Amusement rides* required to comply with 234 shall provide *accessible* routes in accordance with 206.2.9. *Accessible* routes serving *amusement rides* shall comply with Chapter 4 except as modified by 1002.2.

206.2.9.1 Load and Unload Areas. Load and unload areas shall be on an *accessible* route. Where load and unload areas have more than one loading or unloading position, at least one loading and unloading position shall be on an *accessible* route.

206.2.9.2 Wheelchair Spaces, Ride Seats Designed for Transfer, and Transfer Devices. When *amusement rides* are in the load and unload position, *wheelchair spaces* complying with 1002.4, *amusement ride seats* designed for transfer complying with 1002.5, and *transfer devices* complying with 1002.6 shall be on an *accessible* route.

206.2.10 Recreational Boating Facilities. *Boat slips* required to comply with 235.2 and *boarding piers* at *boat launch ramps* required to comply with 235.3 shall be on an *accessible* route. *Accessible* routes serving recreational boating *facilities* shall comply with Chapter 4, except as modified by 1003.2.

206.2.11 Bowling Lanes. Where bowling lanes are provided, at least 5 percent, but no fewer than one of each type of bowling lane, shall be on an *accessible* route.

206.2.12 Court Sports. In court sports, at least one *accessible* route shall directly connect both sides of the court.

206.2.13 Exercise Machines and Equipment. Exercise machines and equipment required to comply with 236 shall be on an *accessible* route.

206.2.14 Fishing Piers and Platforms. Fishing piers and platforms shall be on an *accessible* route. *Accessible* routes serving fishing piers and platforms shall comply with Chapter 4 except as modified by 1005.1.

206.2.15 Golf Facilities. At least one *accessible* route shall connect *accessible elements* and *spaces* within the boundary of the golf course. In addition, *accessible* routes serving golf car rental areas; bag drop areas; course weather shelters complying with 238.2.3; course toilet rooms; and practice putting greens, practice *teeing grounds*, and teeing stations at driving ranges complying with 238.3 shall comply with Chapter 4 except as modified by 1006.2.
 EXCEPTION: *Golf car passages* complying with 1006.3 shall be permitted to be used for all or part of *accessible* routes required by 206.2.15.

206.2.16 Miniature Golf Facilities. Holes required to comply with 239.2, including the start of play, shall be on an *accessible* route. *Accessible* routes serving miniature golf *facilities* shall comply with Chapter 4 except as modified by 1007.2.

206.2.17 Play Areas. *Play areas* shall provide *accessible* routes in accordance with 206.2.17. *Accessible* routes serving *play areas* shall comply with Chapter 4 except as modified by 1008.2.

 206.2.17.1 Ground Level and Elevated Play Components. At least one *accessible* route shall be provided within the *play area*. The *accessible* route shall connect *ground level play components* required to comply with 240.2.1 and *elevated play components* required to comply with 240.2.2, including entry and exit points of the *play components*.

 206.2.17.2 Soft Contained Play Structures. Where three or fewer entry points are provided for *soft contained play structures*, at least one entry point shall be on an *accessible* route. Where four or more entry points are provided for *soft contained play structures*, at least two entry points shall be on an *accessible* route.

206.3 Location. *Accessible* routes shall coincide with or be located in the same area as general *circulation paths*. Where *circulation paths* are interior, required *accessible* routes shall also be interior.

> **Advisory 206.3 Location.** The accessible route must be in the same area as the general circulation path. This means that circulation paths, such as vehicular ways designed for pedestrian traffic, walks, and unpaved paths that are designed to be routinely used by pedestrians must be accessible or have an accessible route nearby. Additionally, accessible vertical interior circulation must be in the same area as stairs and escalators, not isolated in the back of the facility.

206.4 Entrances. *Entrances* shall be provided in accordance with 206.4. *Entrance* doors, doorways, and gates shall comply with 404 and shall be on an *accessible* route complying with 402.
  EXCEPTIONS: 1. Where an *alteration* includes *alterations* to an *entrance*, and the *building* or *facility* has another *entrance* complying with 404 that is on an *accessible* route, the *altered entrance* shall not be required to comply with 206.4 unless required by 202.4.
  2. Where exceptions for *alterations* to *qualified historic buildings or facilities* are permitted by 202.5, no more than one *public entrance* shall be required to comply with 206.4. Where no *public entrance* can comply with 206.4 under criteria established in 202.5 Exception, then either an unlocked *entrance* not used by the public shall comply with 206.4; or a locked *entrance* complying with 206.4 with a notification system or remote monitoring shall be provided.

206.4.1 Public Entrances. In addition to *entrances* required by 206.4.2 through 206.4.9, at least 60 percent of all *public entrances* shall comply with 404.

206.4.2 Parking Structure Entrances. Where direct access is provided for pedestrians from a parking structure to a *building* or *facility entrance*, each direct access to the *building* or *facility entrance* shall comply with 404.

206.4.3 Entrances from Tunnels or Elevated Walkways. Where direct access is provided for pedestrians from a pedestrian tunnel or elevated walkway to a *building* or *facility*, at least one direct *entrance* to the *building* or *facility* from each tunnel or walkway shall comply with 404.

206.4.4 Transportation Facilities. In addition to the requirements of 206.4.2, 206.4.3, and 206.4.5 through 206.4.9, transportation *facilities* shall provide *entrances* in accordance with 206.4.4.

  206.4.4.1 Location. In transportation *facilities*, where different *entrances* serve different transportation fixed routes or groups of fixed routes, at least one *public entrance* serving each fixed route or group of fixed routes shall comply with 404.
    EXCEPTION: *Entrances* to *key stations* and existing intercity rail stations retrofitted in accordance with 49 CFR 37.49 or 49 CFR 37.51 shall not be required to comply with 206.4.4.1.

206.4.4.2 Direct Connections. Direct connections to other *facilities* shall provide an *accessible* route complying with 404 from the point of connection to boarding platforms and all transportation system *elements* required to be *accessible*. Any *elements* provided to facilitate future direct connections shall be on an *accessible* route connecting boarding platforms and all transportation system *elements* required to be *accessible*.

> EXCEPTION: In *key stations* and existing intercity rail stations, existing direct connections shall not be required to comply with 404.

206.4.4.3 Key Stations and Intercity Rail Stations. *Key stations* and existing intercity rail stations required by Subpart C of 49 CFR part 37 to be *altered*, shall have at least one *entrance* complying with 404.

206.4.5 Tenant Spaces. At least one *accessible entrance* to each tenancy in a *facility* shall comply with 404.

> EXCEPTION: *Self-service storage facilities* not required to comply with 225.3 shall not be required to be on an accessible route.

206.4.6 Residential Dwelling Unit Primary Entrance. In *residential dwelling units*, at least one primary *entrance* shall comply with 404. The primary *entrance* to a *residential dwelling unit* shall not be to a bedroom.

206.4.7 Restricted Entrances. Where *restricted entrances* are provided to a *building* or *facility*, at least one *restricted entrance* to the *building* or *facility* shall comply with 404.

206.4.8 Service Entrances. If a *service entrance* is the only *entrance* to a *building* or to a tenancy in a *facility*, that *entrance* shall comply with 404.

206.4.9 Entrances for Inmates or Detainees. Where *entrances* used only by inmates or detainees and security personnel are provided at judicial *facilities,* detention *facilities,* or correctional *facilities*, at least one such *entrance* shall comply with 404.

206.5 Doors, Doorways, and Gates. Doors, doorways, and gates providing user passage shall be provided in accordance with 206.5.

206.5.1 Entrances. Each *entrance* to a *building* or *facility* required to comply with 206.4 shall have at least one door, doorway, or gate complying with 404.

206.5.2 Rooms and Spaces. Within a *building* or *facility*, at least one door, doorway, or gate serving each room or *space* complying with these requirements shall comply with 404.

206.5.3 Transient Lodging Facilities. In *transient lodging facilities*, *entrances*, doors, and doorways providing user passage into and within guest rooms that are not required to provide mobility features complying with 806.2 shall comply with 404.2.3.

> EXCEPTION: Shower and sauna doors in guest rooms that are not required to provide mobility features complying with 806.2 shall not be required to comply with 404.2.3.

206.5.4 Residential Dwelling Units. In *residential dwelling units* required to provide mobility features complying with 809.2 through 809.4, all doors and doorways providing user passage shall comply with 404.

206.6 Elevators. Elevators provided for passengers shall comply with 407. Where multiple elevators are provided, each elevator shall comply with 407.

EXCEPTIONS: 1. In a *building* or *facility* permitted to use the exceptions to 206.2.3 or permitted by 206.7 to use a platform lift, elevators complying with 408 shall be permitted.

2. Elevators complying with 408 or 409 shall be permitted in multi-*story residential dwelling units*.

206.6.1 Existing Elevators. Where *elements* of existing elevators are *altered*, the same *element* shall also be *altered* in all elevators that are programmed to respond to the same hall call control as the *altered* elevator and shall comply with the requirements of 407 for the *altered element*.

206.7 Platform Lifts. Platform lifts shall comply with 410. Platform lifts shall be permitted as a component of an *accessible* route in new construction in accordance with 206.7. Platform lifts shall be permitted as a component of an *accessible* route in an existing *building* or *facility*.

206.7.1 Performance Areas and Speakers' Platforms. Platform lifts shall be permitted to provide *accessible* routes to performance areas and speakers' platforms.

206.7.2 Wheelchair Spaces. Platform lifts shall be permitted to provide an *accessible* route to comply with the *wheelchair space* dispersion and line-of-sight requirements of 221 and 802.

206.7.3 Incidental Spaces. Platform lifts shall be permitted to provide an *accessible* route to incidental *spaces* which are not *public use spaces* and which are occupied by five persons maximum.

206.7.4 Judicial Spaces. Platform lifts shall be permitted to provide an *accessible* route to: jury boxes and witness stands; raised courtroom stations including, judges' benches, clerks' stations, bailiffs' stations, deputy clerks' stations, and court reporters' stations; and to depressed areas such as the well of a court.

206.7.5 Existing Site Constraints. Platform lifts shall be permitted where existing exterior *site* constraints make use of a *ramp* or elevator infeasible.

> **Advisory 206.7.5 Existing Site Constraints.** This exception applies where topography or other similar existing site constraints necessitate the use of a platform lift as the only feasible alternative. While the site constraint must reflect exterior conditions, the lift can be installed in the interior of a building. For example, a new building constructed between and connected to two existing buildings may have insufficient space to coordinate floor levels and also to provide ramped entry from the public way. In this example, an exterior or interior platform lift could be used to provide an accessible entrance or to coordinate one or more interior floor levels.

206.7.6 Guest Rooms and Residential Dwelling Units. Platform lifts shall be permitted to connect levels within *transient lodging* guest rooms required to provide mobility features complying with 806.2 or *residential dwelling units* required to provide mobility features complying with 809.2 through 809.4.

206.7.7 Amusement Rides. Platform lifts shall be permitted to provide *accessible* routes to load and unload areas serving *amusement rides*.

206.7.8 Play Areas. Platform lifts shall be permitted to provide *accessible* routes to *play components* or *soft contained play structures*.

206.7.9 Team or Player Seating. Platform lifts shall be permitted to provide *accessible* routes to team or player seating areas serving *areas of sport activity*.

> **Advisory 206.7.9 Team or Player Seating.** While the use of platform lifts is allowed, ramps are recommended to provide access to player seating areas serving an area of sport activity.

206.7.10 Recreational Boating Facilities and Fishing Piers and Platforms. Platform lifts shall be permitted to be used instead of *gangways* that are part of *accessible* routes serving recreational boating *facilities* and fishing piers and platforms.

206.8 Security Barriers. Security barriers, including but not limited to, security bollards and security check points, shall not obstruct a required *accessible* route or *accessible means of egress*.

EXCEPTION: Where security barriers incorporate *elements* that cannot comply with these requirements such as certain metal detectors, fluoroscopes, or other similar devices, the *accessible* route shall be permitted to be located adjacent to security screening devices. The *accessible* route shall permit persons with disabilities passing around security barriers to maintain visual contact with their personal items to the same extent provided others passing through the security barrier.

207 Accessible Means of Egress

207.1 General. Means of egress shall comply with section 1003.2.13 of the International Building Code (2000 edition and 2001 Supplement) or section 1007 of the International Building Code (2003 edition) (incorporated by reference, see "Referenced Standards" in Chapter 1).

EXCEPTIONS: 1. Where means of egress are permitted by local *building* or life safety codes to share a common path of egress travel, *accessible means of egress* shall be permitted to share a common path of egress travel.

2. Areas of refuge shall not be required in detention and correctional *facilities*.

207.2 Platform Lifts. Standby power shall be provided for platform lifts permitted by section 1003.2.13.4 of the International Building Code (2000 edition and 2001 Supplement) or section 1007.5 of the International Building Code (2003 edition) (incorporated by reference, see "Referenced Standards" in Chapter 1) to serve as a part of an *accessible means of egress*.

208 Parking Spaces

208.1 General. Where parking *spaces* are provided, parking *spaces* shall be provided in accordance with 208.

EXCEPTION: Parking *spaces* used exclusively for buses, trucks, other delivery vehicles, law enforcement vehicles, or vehicular impound shall not be required to comply with 208 provided that lots accessed by the public are provided with a passenger loading zone complying with 503.

208.2 Minimum Number. Parking *spaces* complying with 502 shall be provided in accordance with Table 208.2 except as required by 208.2.1, 208.2.2, and 208.2.3. Where more than one parking *facility* is provided on a *site*, the number of *accessible spaces* provided on the *site* shall be calculated according to the number of *spaces* required for each parking *facility*.

Table 208.2 Parking Spaces

Total Number of Parking Spaces Provided in Parking Facility	Minimum Number of Required Accessible Parking Spaces
1 to 25	1
26 to 50	2
51 to 75	3
76 to 100	4
101 to 150	5
151 to 200	6
201 to 300	7
301 to 400	8
401 to 500	9
501 to 1000	2 percent of total
1001 and over	20, plus 1 for each 100, or fraction thereof, over 1000

Advisory 208.2 Minimum Number. The term "parking facility" is used Section 208.2 instead of the term "parking lot" so that it is clear that both parking lots and parking structures are required to comply with this section. The number of parking spaces required to be accessible is to be calculated separately for each parking facility; the required number is not to be based on the total number of parking spaces provided in all of the parking facilities provided on the site.

208.2.1 Hospital Outpatient Facilities. Ten percent of patient and visitor parking *spaces* provided to serve hospital outpatient *facilities* shall comply with 502.

> **Advisory 208.2.1 Hospital Outpatient Facilities.** The term "outpatient facility" is not defined in this document but is intended to cover facilities or units that are located in hospitals and that provide regular and continuing medical treatment without an overnight stay. Doctors' offices, independent clinics, or other facilities not located in hospitals are not considered hospital outpatient facilities for purposes of this document.

208.2.2 Rehabilitation Facilities and Outpatient Physical Therapy Facilities. Twenty percent of patient and visitor parking *spaces* provided to serve rehabilitation *facilities* specializing in treating conditions that affect mobility and outpatient physical therapy *facilities* shall comply with 502.

> **Advisory 208.2.2 Rehabilitation Facilities and Outpatient Physical Therapy Facilities.** Conditions that affect mobility include conditions requiring the use or assistance of a brace, cane, crutch, prosthetic device, wheelchair, or powered mobility aid; arthritic, neurological, or orthopedic conditions that severely limit one's ability to walk; respiratory diseases and other conditions which may require the use of portable oxygen; and cardiac conditions that impose significant functional limitations.

208.2.3 Residential Facilities. Parking *spaces* provided to serve residential *facilities* shall comply with 208.2.3.

208.2.3.1 Parking for Residents. Where at least one parking *space* is provided for each *residential dwelling unit*, at least one parking *space* complying with 502 shall be provided for each *residential dwelling unit* required to provide mobility features complying with 809.2 through 809.4.

208.2.3.2 Additional Parking Spaces for Residents. Where the total number of parking *spaces* provided for each *residential dwelling unit* exceeds one parking *space* per *residential dwelling unit*, 2 percent, but no fewer than one *space*, of all the parking *spaces* not covered by 208.2.3.1 shall comply with 502.

208.2.3.3 Parking for Guests, Employees, and Other Non-Residents. Where parking spaces are provided for persons other than residents, parking shall be provided in accordance with Table 208.2.

208.2.4 Van Parking Spaces. For every six or fraction of six parking *spaces* required by 208.2 to comply with 502, at least one shall be a van parking *space* complying with 502.

208.3 Location. Parking *facilities* shall comply with 208.3

208.3.1 General. Parking *spaces* complying with 502 that serve a particular *building* or *facility* shall be located on the shortest *accessible* route from parking to an *entrance* complying with 206.4. Where parking serves more than one *accessible entrance*, parking *spaces* complying with 502 shall be dispersed and located on the shortest *accessible* route to the *accessible entrances*. In parking

facilities that do not serve a particular *building* or *facility*, parking *spaces* complying with 502 shall be located on the shortest *accessible* route to an *accessible* pedestrian *entrance* of the parking *facility*.
EXCEPTIONS: 1. All van parking *spaces* shall be permitted to be grouped on one level within a multi-*story* parking *facility*.
2. Parking *spaces* shall be permitted to be located in different parking *facilities* if substantially equivalent or greater *accessibility* is provided in terms of distance from an *accessible entrance* or *entrances*, parking fee, and user convenience.

> **Advisory 208.3.1 General Exception 2.** Factors that could affect "user convenience" include, but are not limited to, protection from the weather, security, lighting, and comparative maintenance of the alternative parking site.

208.3.2 Residential Facilities. In residential *facilities* containing *residential dwelling units* required to provide mobility features complying with 809.2 through 809.4, parking *spaces* provided in accordance with 208.2.3.1 shall be located on the shortest *accessible* route to the *residential dwelling unit entrance* they serve. *Spaces* provided in accordance with 208.2.3.2 shall be dispersed throughout all types of parking provided for the *residential dwelling units*.
EXCEPTION: Parking *spaces* provided in accordance with 208.2.3.2 shall not be required to be dispersed throughout all types of parking if substantially equivalent or greater *accessibility* is provided in terms of distance from an *accessible entrance*, parking fee, and user convenience.

> **Advisory 208.3.2 Residential Facilities Exception.** Factors that could affect "user convenience" include, but are not limited to, protection from the weather, security, lighting, and comparative maintenance of the alternative parking site.

209 Passenger Loading Zones and Bus Stops

209.1 General. Passenger loading zones shall be provided in accordance with 209.

209.2 Type. Where provided, passenger loading zones shall comply with 209.2.

209.2.1 Passenger Loading Zones. Passenger loading zones, except those required to comply with 209.2.2 and 209.2.3, shall provide at least one passenger loading zone complying with 503 in every continuous 100 linear feet (30 m) of loading zone *space*, or fraction thereof.

209.2.2 Bus Loading Zones. In bus loading zones restricted to use by designated or specified public transportation vehicles, each bus bay, bus stop, or other area designated for lift or *ramp* deployment shall comply with 810.2.

> **Advisory 209.2.2 Bus Loading Zones.** The terms "designated public transportation" and "specified public transportation" are defined by the Department of Transportation at 49 CFR 37.3 in regulations implementing the Americans with Disabilities Act. These terms refer to public transportation services provided by public or private entities, respectively. For example, designated public transportation vehicles include buses and vans operated by public transit agencies, while specified public transportation vehicles include tour and charter buses, taxis and limousines, and hotel shuttles operated by private entities.

209.2.3 On-Street Bus Stops. On-street bus stops shall comply with 810.2 to the maximum extent practicable.

209.3 Medical Care and Long-Term Care Facilities. At least one passenger loading zone complying with 503 shall be provided at an *accessible entrance* to licensed medical care and licensed long-term care *facilities* where the period of stay exceeds twenty-four hours.

209.4 Valet Parking. Parking *facilities* that provide valet parking services shall provide at least one passenger loading zone complying with 503.

209.5 Mechanical Access Parking Garages. Mechanical access parking garages shall provide at least one passenger loading zone complying with 503 at vehicle drop-off and vehicle pick-up areas.

210 Stairways

210.1 General. Interior and exterior stairs that are part of a means of egress shall comply with 504.
    EXCEPTIONS: 1. In detention and correctional *facilities*, stairs that are not located in *public use* areas shall not be required to comply with 504.
    2. In *alterations*, stairs between levels that are connected by an *accessible* route shall not be required to comply with 504, except that handrails complying with 505 shall be provided when the stairs are *altered*.
    3. In *assembly areas*, aisle stairs shall not be required to comply with 504.
    4. Stairs that connect *play components* shall not be required to comply with 504.

> **Advisory 210.1 General.** Although these requirements do not mandate handrails on stairs that are not part of a means of egress, State or local building codes may require handrails or guards.

211 Drinking Fountains

211.1 General. Where drinking fountains are provided on an exterior *site*, on a floor, or within a secured area they shall be provided in accordance with 211.
    EXCEPTION: In detention or correctional *facilities*, drinking fountains only serving holding or housing cells not required to comply with 232 shall not be required to comply with 211.

211.2 Minimum Number. No fewer than two drinking fountains shall be provided. One drinking fountain shall comply with 602.1 through 602.6 and one drinking fountain shall comply with 602.7.
    EXCEPTION: Where a single drinking fountain complies with 602.1 through 602.6 and 602.7, it shall be permitted to be substituted for two separate drinking fountains.

211.3 More Than Minimum Number. Where more than the minimum number of drinking fountains specified in 211.2 are provided, 50 percent of the total number of drinking fountains provided shall comply with 602.1 through 602.6, and 50 percent of the total number of drinking fountains provided shall comply with 602.7.

EXCEPTION: Where 50 percent of the drinking fountains yields a fraction, 50 percent shall be permitted to be rounded up or down provided that the total number of drinking fountains complying with 211 equals 100 percent of drinking fountains.

212 Kitchens, Kitchenettes, and Sinks

212.1 General. Where provided, kitchens, kitchenettes, and sinks shall comply with 212.

212.2 Kitchens and Kitchenettes. Kitchens and kitchenettes shall comply with 804.

212.3 Sinks. Where sinks are provided, at least 5 percent, but no fewer than one, of each type provided in each *accessible* room or *space* shall comply with 606.
EXCEPTION: Mop or service sinks shall not be required to comply with 212.3.

213 Toilet Facilities and Bathing Facilities

213.1 General. Where toilet *facilities* and bathing *facilities* are provided, they shall comply with 213. Where toilet *facilities* and bathing *facilities* are provided in *facilities* permitted by 206.2.3 Exceptions 1 and 2 not to connect *stories* by an *accessible* route, toilet *facilities* and bathing *facilities* shall be provided on a *story* connected by an *accessible* route to an *accessible* entrance.

213.2 Toilet Rooms and Bathing Rooms. Where toilet rooms are provided, each toilet room shall comply with 603. Where bathing rooms are provided, each bathing room shall comply with 603.
EXCEPTIONS: 1. In *alterations* where it is *technically infeasible* to comply with 603, altering existing toilet or bathing rooms shall not be required where a single unisex toilet room or bathing room complying with 213.2.1 is provided and located in the same area and on the same floor as existing inaccessible toilet or bathing rooms.
2. Where exceptions for *alterations* to *qualified historic buildings or facilities* are permitted by 202.5, no fewer than one toilet room for each sex complying with 603 or one unisex toilet room complying with 213.2.1 shall be provided.
3. Where multiple single user portable toilet or bathing units are clustered at a single location, no more than 5 percent of the toilet units and bathing units at each cluster shall be required to comply with 603. Portable toilet units and bathing units complying with 603 shall be identified by the International Symbol of *Accessibility* complying with 703.7.2.1.
4. Where multiple single user toilet rooms are clustered at a single location, no more than 50 percent of the single user toilet rooms for each use at each cluster shall be required to comply with 603.

> **Advisory 213.2 Toilet Rooms and Bathing Rooms.** These requirements allow the use of unisex (or single-user) toilet rooms in alterations when technical infeasibility can be demonstrated. Unisex toilet rooms benefit people who use opposite sex personal care assistants. For this reason, it is advantageous to install unisex toilet rooms in addition to accessible single-sex toilet rooms in new facilities.
>
> **Advisory 213.2 Toilet Rooms and Bathing Rooms Exceptions 3 and 4.** A "cluster" is a group of toilet rooms proximate to one another. Generally, toilet rooms in a cluster are within sight of, or adjacent to, one another.

213.2.1 Unisex (Single-Use or Family) Toilet and Unisex Bathing Rooms. Unisex toilet rooms shall contain not more than one lavatory, and two water closets without urinals or one water closet and one urinal. Unisex bathing rooms shall contain one shower or one shower and one bathtub, one lavatory, and one water closet. Doors to unisex toilet rooms and unisex bathing rooms shall have privacy latches.

213.3 Plumbing Fixtures and Accessories. Plumbing fixtures and accessories provided in a toilet room or bathing room required to comply with 213.2 shall comply with 213.3.

213.3.1 Toilet Compartments. Where toilet compartments are provided, at least one toilet compartment shall comply with 604.8.1. In addition to the compartment required to comply with 604.8.1, at least one compartment shall comply with 604.8.2 where six or more toilet compartments are provided, or where the combination of urinals and water closets totals six or more fixtures.

> **Advisory 213.3.1 Toilet Compartments.** A toilet compartment is a partitioned space that is located within a toilet room, and that normally contains no more than one water closet. A toilet compartment may also contain a lavatory. A lavatory is a sink provided for hand washing. Full-height partitions and door assemblies can comprise toilet compartments where the minimum required spaces are provided within the compartment.

213.3.2 Water Closets. Where water closets are provided, at least one shall comply with 604.

213.3.3 Urinals. Where more than one urinal is provided, at least one shall comply with 605.

213.3.4 Lavatories. Where lavatories are provided, at least one shall comply with 606 and shall not be located in a toilet compartment.

213.3.5 Mirrors. Where mirrors are provided, at least one shall comply with 603.3.

213.3.6 Bathing Facilities. Where bathtubs or showers are provided, at least one bathtub complying with 607 or at least one shower complying with 608 shall be provided.

213.3.7 Coat Hooks and Shelves. Where coat hooks or shelves are provided in toilet rooms without toilet compartments, at least one of each type shall comply with 603.4. Where coat hooks or shelves are provided in toilet compartments, at least one of each type complying with 604.8.3 shall be provided in toilet compartments required to comply with 213.3.1. Where coat hooks or shelves are provided in bathing *facilities*, at least one of each type complying with 603.4 shall serve fixtures required to comply with 213.3.6.

214 Washing Machines and Clothes Dryers

214.1 General. Where provided, washing machines and clothes dryers shall comply with 214.

214.2 Washing Machines. Where three or fewer washing machines are provided, at least one shall comply with 611. Where more than three washing machines are provided, at least two shall comply with 611.

214.3 Clothes Dryers. Where three or fewer clothes dryers are provided, at least one shall comply with 611. Where more than three clothes dryers are provided, at least two shall comply with 611.

215 Fire Alarm Systems

215.1 General. Where fire alarm systems provide audible alarm coverage, alarms shall comply with 215.

EXCEPTION: In existing *facilities*, visible alarms shall not be required except where an existing fire alarm system is upgraded or replaced, or a new fire alarm system is installed.

> **Advisory 215.1 General.** Unlike audible alarms, visible alarms must be located within the space they serve so that the signal is visible. Facility alarm systems (other than fire alarm systems) such as those used for tornado warnings and other emergencies are not required to comply with the technical criteria for alarms in Section 702. Every effort should be made to ensure that such alarms can be differentiated in their signal from fire alarms systems and that people who need to be notified of emergencies are adequately safeguarded. Consult local fire departments and prepare evacuation plans taking into consideration the needs of every building occupant, including people with disabilities.

215.2 Public and Common Use Areas. Alarms in *public use* areas and *common use* areas shall comply with 702.

215.3 Employee Work Areas. Where *employee work areas* have audible alarm coverage, the wiring system shall be designed so that visible alarms complying with 702 can be integrated into the alarm system.

215.4 Transient Lodging. Guest rooms required to comply with 224.4 shall provide alarms complying with 702.

215.5 Residential Facilities. Where provided in *residential dwelling units* required to comply with 809.5, alarms shall comply with 702.

216 Signs

216.1 General. Signs shall be provided in accordance with 216 and shall comply with 703.
EXCEPTIONS: 1. *Building* directories, menus, seat and row designations in *assembly areas*, occupant names, *building* addresses, and company names and logos shall not be required to comply with 216.
2. In parking *facilities*, signs shall not be required to comply with 216.2, 216.3, and 216.6 through 216.12.
3. Temporary, 7 days or less, signs shall not be required to comply with 216.
4. In detention and correctional *facilities*, signs not located in *public use* areas shall not be required to comply with 216.

216.2 Designations. Interior and exterior signs identifying permanent rooms and *spaces* shall comply with 703.1, 703.2, and 703.5. Where *pictograms* are provided as designations of permanent interior rooms and *spaces*, the *pictograms* shall comply with 703.6 and shall have text descriptors complying with 703.2 and 703.5.

EXCEPTION: Exterior signs that are not located at the door to the *space* they serve shall not be required to comply with 703.2.

> **Advisory 216.2 Designations.** Section 216.2 applies to signs that provide designations, labels, or names for interior rooms or spaces where the sign is not likely to change over time. Examples include interior signs labeling restrooms, room and floor numbers or letters, and room names. Tactile text descriptors are required for pictograms that are provided to label or identify a permanent room or space. Pictograms that provide information about a room or space, such as "no smoking," occupant logos, and the International Symbol of Accessibility, are not required to have text descriptors.

216.3 Directional and Informational Signs. Signs that provide direction to or information about interior *spaces* and *facilities* of the *site* shall comply with 703.5.

> **Advisory 216.3 Directional and Informational Signs.** Information about interior spaces and facilities includes rules of conduct, occupant load, and similar signs. Signs providing direction to rooms or spaces include those that identify egress routes.

216.4 Means of Egress. Signs for means of egress shall comply with 216.4.

216.4.1 Exit Doors. Doors at exit passageways, exit discharge, and exit stairways shall be identified by *tactile* signs complying with 703.1, 703.2, and 703.5.

> **Advisory 216.4.1 Exit Doors.** An exit passageway is a horizontal exit component that is separated from the interior spaces of the building by fire-resistance-rated construction and that leads to the exit discharge or public way. The exit discharge is that portion of an egress system between the termination of an exit and a public way.

216.4.2 Areas of Refuge. Signs required by section 1003.2.13.5.4 of the International Building Code (2000 edition) or section 1007.6.4 of the International Building Code (2003 edition) (incorporated by reference, see "Referenced Standards" in Chapter 1) to provide instructions in areas of refuge shall comply with 703.5.

216.4.3 Directional Signs. Signs required by section 1003.2.13.6 of the International Building Code (2000 edition) or section 1007.7 of the International Building Code (2003 edition) (incorporated by reference, see "Referenced Standards" in Chapter 1) to provide directions to *accessible means of egress* shall comply with 703.5.

216.5 Parking. Parking *spaces* complying with 502 shall be identified by signs complying with 502.6.
EXCEPTIONS: 1. Where a total of four or fewer parking *spaces*, including *accessible* parking *spaces*, are provided on a *site*, identification of *accessible* parking *spaces* shall not be required.
2. In residential *facilities*, where parking *spaces* are assigned to specific *residential dwelling units*, identification of *accessible* parking *spaces* shall not be required.

216.6 Entrances. Where not all *entrances* comply with 404, *entrances* complying with 404 shall be identified by the International Symbol of *Accessibility* complying with 703.7.2.1. Directional signs complying with 703.5 that indicate the location of the nearest *entrance* complying with 404 shall be provided at *entrances* that do not comply with 404.

> **Advisory 216.6 Entrances.** Where a directional sign is required, it should be located to minimize backtracking. In some cases, this could mean locating a sign at the beginning of a route, not just at the inaccessible entrances to a building.

216.7 Elevators. Where existing elevators do not comply with 407, elevators complying with 407 shall be clearly identified with the International Symbol of *Accessibility* complying with 703.7.2.1.

216.8 Toilet Rooms and Bathing Rooms. Where existing toilet rooms or bathing rooms do not comply with 603, directional signs indicating the location of the nearest toilet room or bathing room complying with 603 within the *facility* shall be provided. Signs shall comply with 703.5 and shall include the International Symbol of *Accessibility* complying with 703.7.2.1. Where existing toilet rooms or bathing rooms do not comply with 603, the toilet rooms or bathing rooms complying with 603 shall be identified by the International Symbol of *Accessibility* complying with 703.7.2.1. Where clustered single user toilet rooms or bathing *facilities* are permitted to use exceptions to 213.2, toilet rooms or bathing *facilities* complying with 603 shall be identified by the International Symbol of *Accessibility* complying with 703.7.2.1 unless all toilet rooms and bathing *facilities* comply with 603.

216.9 TTYs. Identification and directional signs for public *TTYs* shall be provided in accordance with 216.9.

216.9.1 Identification Signs. Public *TTYs* shall be identified by the International Symbol of *TTY* complying with 703.7.2.2.

216.9.2 Directional Signs. Directional signs indicating the location of the nearest public *TTY* shall be provided at all banks of public pay telephones not containing a public *TTY*. In addition, where signs provide direction to public pay telephones, they shall also provide direction to public *TTYs*. Directional signs shall comply with 703.5 and shall include the International Symbol of *TTY* complying with 703.7.2.2.

216.10 Assistive Listening Systems. Each *assembly area* required by 219 to provide *assistive listening systems* shall provide signs informing patrons of the availability of the *assistive listening system*. Assistive listening signs shall comply with 703.5 and shall include the International Symbol of Access for Hearing Loss complying with 703.7.2.4.
EXCEPTION: Where ticket offices or windows are provided, signs shall not be required at each *assembly area* provided that signs are displayed at each ticket office or window informing patrons of the availability of *assistive listening systems*.

216.11 Check-Out Aisles. Where more than one check-out aisle is provided, check-out aisles complying with 904.3 shall be identified by the International Symbol of *Accessibility* complying with 703.7.2.1. Where check-out aisles are identified by numbers, letters, or functions, signs identifying

check-out aisles complying with 904.3 shall be located in the same location as the check-out aisle identification.
>   **EXCEPTION:** Where all check-out aisles serving a single function comply with 904.3, signs complying with 703.7.2.1 shall not be required.

216.12 Amusement Rides. Signs identifying the type of access provided on *amusement rides* shall be provided at entries to queues and waiting lines. In addition, where *accessible* unload areas also serve as *accessible* load areas, signs indicating the location of the *accessible* load and unload areas shall be provided at entries to queues and waiting lines.

> **Advisory 216.12 Amusement Rides.** Amusement rides designed primarily for children, amusement rides that are controlled or operated by the rider, and amusement rides without seats, are not required to provide wheelchair spaces, transfer seats, or transfer systems, and need not meet the sign requirements in 216.12. The load and unload areas of these rides must, however, be on an accessible route and must provide turning space.

217 Telephones

217.1 General. Where coin-operated public pay telephones, coinless public pay telephones, public *closed-circuit telephones*, public courtesy phones, or other types of public telephones are provided, public telephones shall be provided in accordance with 217 for each type of public telephone provided. For purposes of this section, a bank of telephones shall be considered to be two or more adjacent telephones.

> **Advisory 217.1 General.** These requirements apply to all types of public telephones including courtesy phones at airports and rail stations that provide a free direct connection to hotels, transportation services, and tourist attractions.

217.2 Wheelchair Accessible Telephones. Where public telephones are provided, wheelchair *accessible* telephones complying with 704.2 shall be provided in accordance with Table 217.2.
>   **EXCEPTION:** Drive-up only public telephones shall not be required to comply with 217.2.

Table 217.2 Wheelchair Accessible Telephones

Number of Telephones Provided on a Floor, Level, or Exterior Site	Minimum Number of Required Wheelchair Accessible Telephones
1 or more single units	1 per floor, level, and exterior *site*
1 bank	1 per floor, level, and exterior *site*
2 or more banks	1 per bank

217.3 Volume Controls. All public telephones shall have volume controls complying with 704.3.

217.4 TTYs. *TTYs* complying with 704.4 shall be provided in accordance with 217.4.

> **Advisory 217.4 TTYs.** Separate requirements are provided based on the number of public pay telephones provided at a bank of telephones, within a floor, a building, or on a site. In some instances one TTY can be used to satisfy more than one of these requirements. For example, a TTY required for a bank can satisfy the requirements for a building. However, the requirement for at least one TTY on an exterior site cannot be met by installing a TTY in a bank inside a building. Consideration should be given to phone systems that can accommodate both digital and analog transmissions for compatibility with digital and analog TTYs.

217.4.1 Bank Requirement. Where four or more public pay telephones are provided at a bank of telephones, at least one public *TTY* complying with 704.4 shall be provided at that bank.

   EXCEPTION: *TTYs* shall not be required at banks of telephones located within 200 feet (61 m) of, and on the same floor as, a bank containing a public *TTY*.

217.4.2 Floor Requirement. *TTYs* in *public buildings* shall be provided in accordance with 217.4.2.1. *TTYs* in *private buildings* shall be provided in accordance with 217.4.2.2.

   217.4.2.1 Public Buildings. Where at least one public pay telephone is provided on a floor of a *public building*, at least one public *TTY* shall be provided on that floor.

   217.4.2.2 Private Buildings. Where four or more public pay telephones are provided on a floor of a *private building*, at least one public *TTY* shall be provided on that floor.

217.4.3 Building Requirement. *TTYs* in *public buildings* shall be provided in accordance with 217.4.3.1. *TTYs* in *private buildings* shall be provided in accordance with 217.4.3.2.

   217.4.3.1 Public Buildings. Where at least one public pay telephone is provided in a *public building*, at least one public *TTY* shall be provided in the *building*. Where at least one public pay telephone is provided in a *public use* area of a *public building*, at least one public *TTY* shall be provided in the *public building* in a *public use* area.

   217.4.3.2 Private Buildings. Where four or more public pay telephones are provided in a *private building*, at least one public *TTY* shall be provided in the *building*.

217.4.4 Exterior Site Requirement. Where four or more public pay telephones are provided on an exterior *site*, at least one public *TTY* shall be provided on the *site*.

217.4.5 Rest Stops, Emergency Roadside Stops, and Service Plazas. Where at least one public pay telephone is provided at a public rest stop, emergency roadside stop, or service plaza, at least one public *TTY* shall be provided.

217.4.6 Hospitals. Where at least one public pay telephone is provided serving a hospital emergency room, hospital recovery room, or hospital waiting room, at least one public *TTY* shall be provided at each location.

217.4.7 Transportation Facilities. In transportation *facilities*, in addition to the requirements of 217.4.1 through 217.4.4, where at least one public pay telephone serves a particular *entrance* to a bus or rail *facility*, at least one public *TTY* shall be provided to serve that *entrance*. In airports, in addition to the requirements of 217.4.1 through 217.4.4, where four or more public pay telephones are located in a terminal outside the security areas, a concourse within the security areas, or a baggage claim area in a terminal, at least one public *TTY* shall be provided in each location.

217.4.8 Detention and Correctional Facilities. In detention and correctional *facilities*, where at least one pay telephone is provided in a secured area used only by detainees or inmates and security personnel, at least one *TTY* shall be provided in at least one secured area.

217.5 Shelves for Portable TTYs. Where a bank of telephones in the interior of a *building* consists of three or more public pay telephones, at least one public pay telephone at the bank shall be provided with a shelf and an electrical outlet in accordance with 704.5.

EXCEPTIONS: 1. Secured areas of detention and correctional *facilities* where shelves and outlets are prohibited for purposes of security or safety shall not be required to comply with 217.5.

2. The shelf and electrical outlet shall not be required at a bank of telephones with a *TTY*.

218 Transportation Facilities

218.1 General. Transportation *facilities* shall comply with 218.

218.2 New and Altered Fixed Guideway Stations. New and *altered* stations in rapid rail, light rail, commuter rail, intercity rail, high speed rail, and other fixed guideway systems shall comply with 810.5 through 810.10.

218.3 Key Stations and Existing Intercity Rail Stations. *Key stations* and existing intercity rail stations shall comply with 810.5 through 810.10.

218.4 Bus Shelters. Where provided, bus shelters shall comply with 810.3.

218.5 Other Transportation Facilities. In other transportation *facilities*, public address systems shall comply with 810.7 and clocks shall comply with 810.8.

219 Assistive Listening Systems

219.1 General. *Assistive listening systems* shall be provided in accordance with 219 and shall comply with 706.

219.2 Required Systems. In each *assembly area* where audible communication is integral to the use of the *space*, an *assistive listening system* shall be provided.

EXCEPTION: Other than in courtrooms, *assistive listening systems* shall not be required where audio amplification is not provided.

219.3 Receivers. Receivers complying with 706.2 shall be provided for *assistive listening systems* in each *assembly area* in accordance with Table 219.3. Twenty-five percent minimum of receivers provided, but no fewer than two, shall be hearing-aid compatible in accordance with 706.3.

EXCEPTIONS: 1. Where a *building* contains more than one *assembly area* and the *assembly areas* required to provide *assistive listening systems* are under one management, the total number of required receivers shall be permitted to be calculated according to the total number of seats in the *assembly areas* in the *building* provided that all receivers are usable with all systems.
2. Where all seats in an *assembly area* are served by an induction loop *assistive listening system*, the minimum number of receivers required by Table 219.3 to be hearing-aid compatible shall not be required to be provided.

Table 219.3 Receivers for Assistive Listening Systems

Capacity of Seating in Assembly Area	Minimum Number of Required Receivers	Minimum Number of Required Receivers Required to be Hearing-aid Compatible
50 or less	2	2
51 to 200	2, plus 1 per 25 seats over 50 seats[1]	2
201 to 500	2, plus 1 per 25 seats over 50 seats[1]	1 per 4 receivers[1]
501 to 1000	20, plus 1 per 33 seats over 500 seats[1]	1 per 4 receivers[1]
1001 to 2000	35, plus 1 per 50 seats over 1000 seats[1]	1 per 4 receivers[1]
2001 and over	55 plus 1 per 100 seats over 2000 seats[1]	1 per 4 receivers[1]

1. Or fraction thereof.

220 Automatic Teller Machines and Fare Machines

220.1 General. Where automatic teller machines or self-service fare vending, collection, or adjustment machines are provided, at least one of each type provided at each location shall comply with 707. Where bins are provided for envelopes, waste paper, or other purposes, at least one of each type shall comply with 811.

> **Advisory 220.1 General.** If a bank provides both interior and exterior ATMs, each such installation is considered a separate location. Accessible ATMs, including those with speech and those that are within reach of people who use wheelchairs, must provide all the functions provided to customers at that location at all times. For example, it is unacceptable for the accessible ATM only to provide cash withdrawals while inaccessible ATMs also sell theater tickets.

221 Assembly Areas

221.1 General. *Assembly areas* shall provide *wheelchair spaces*, companion seats, and designated aisle seats complying with 221 and 802. In addition, lawn seating shall comply with 221.5.
[See additional requirements at 28 CFR 35.151(g), p. 12, and 28 CFR 36.406(f), p. 29.]

221.2 Wheelchair Spaces. *Wheelchair spaces* complying with 221.2 shall be provided in *assembly areas* with fixed seating.

221.2.1 Number and Location. *Wheelchair spaces* shall be provided complying with 221.2.1.

221.2.1.1 General Seating. *Wheelchair spaces* complying with 802.1 shall be provided in accordance with Table 221.2.1.1.

Table 221.2.1.1 Number of Wheelchair Spaces in Assembly Areas

Number of Seats	Minimum Number of Required Wheelchair Spaces
4 to 25	1
26 to 50	2
51 to 150	4
151 to 300	5
301 to 500	6
501 to 5000	6, plus 1 for each 150, or fraction thereof, between 501 through 5000
5001 and over	36, plus 1 for each 200, or fraction thereof, over 5000

221.2.1.2 Luxury Boxes, Club Boxes, and Suites in Arenas, Stadiums, and Grandstands. In each luxury box, club box, and suite within arenas, stadiums, and grandstands, *wheelchair spaces* complying with 802.1 shall be provided in accordance with Table 221.2.1.1.

> **Advisory 221.2.1.2 Luxury Boxes, Club Boxes, and Suites in Arenas, Stadiums, and Grandstands.** The number of wheelchair spaces required in luxury boxes, club boxes, and suites within an arena, stadium, or grandstand is to be calculated box by box and suite by suite.

221.2.1.3 Other Boxes. In boxes other than those required to comply with 221.2.1.2, the total number of *wheelchair spaces* required shall be determined in accordance with Table 221.2.1.1. *Wheelchair spaces* shall be located in not less than 20 percent of all boxes provided. *Wheelchair spaces* shall comply with 802.1.

> **Advisory 221.2.1.3 Other Boxes.** The provision for seating in "other boxes" includes box seating provided in facilities such as performing arts auditoria where tiered boxes are designed for spatial and acoustical purposes. The number of wheelchair spaces required in boxes covered by 221.2.1.3 is calculated based on the total number of seats provided in these other boxes. The resulting number of wheelchair spaces must be located in no fewer than 20% of the boxes covered by this section. For example, a concert hall has 20 boxes, each of which contains 10 seats, totaling 200 seats. In this example, 5 wheelchair spaces would be required, and they must be placed in at least 4 of the boxes. Additionally, because the wheelchair spaces must also meet the dispersion requirements of 221.2.3, the boxes containing these wheelchair spaces cannot all be located in one area unless an exception to the dispersion requirements applies.

221.2.1.4 Team or Player Seating. At least one *wheelchair space* complying with 802.1 shall be provided in team or player seating areas serving *areas of sport activity*.
    EXCEPTION: *Wheelchair spaces* shall not be required in team or player seating areas serving bowling lanes not required to comply with 206.2.11.

221.2.2 Integration. *Wheelchair spaces* shall be an integral part of the seating plan.

> **Advisory 221.2.2 Integration.** The requirement that wheelchair spaces be an "integral part of the seating plan" means that wheelchair spaces must be placed within the footprint of the seating area. Wheelchair spaces cannot be segregated from seating areas. For example, it would be unacceptable to place only the wheelchair spaces, or only the wheelchair spaces and their associated companion seats, outside the seating areas defined by risers in an assembly area.

221.2.3 Lines of Sight and Dispersion. *Wheelchair spaces* shall provide lines of sight complying with 802.2 and shall comply with 221.2.3. In providing lines of sight, *wheelchair spaces* shall be dispersed. *Wheelchair spaces* shall provide spectators with choices of seating locations and viewing angles that are substantially equivalent to, or better than, the choices of seating locations and viewing angles available to all other spectators. When the number of *wheelchair spaces* required by 221.2.1 has been met, further dispersion shall not be required.
    EXCEPTION: *Wheelchair spaces* in team or player seating areas serving *areas of sport activity* shall not be required to comply with 221.2.3.

> **Advisory 221.2.3 Lines of Sight and Dispersion.** Consistent with the overall intent of the ADA, individuals who use wheelchairs must be provided equal access so that their experience is substantially equivalent to that of other members of the audience. Thus, while individuals who use wheelchairs need not be provided with the best seats in the house, neither may they be relegated to the worst.

221.2.3.1 Horizontal Dispersion. *Wheelchair spaces* shall be dispersed horizontally.
    EXCEPTIONS: 1. Horizontal dispersion shall not be required in *assembly areas* with 300 or fewer seats if the companion seats required by 221.3 and *wheelchair spaces* are located within the 2nd or 3rd quartile of the total row length. Intermediate aisles shall be included in

determining the total row length. If the row length in the 2nd and 3rd quartile of a row is insufficient to accommodate the required number of companion seats and *wheelchair spaces*, the additional companion seats and *wheelchair spaces* shall be permitted to be located in the 1st and 4th quartile of the row.
2. In row seating, two *wheelchair spaces* shall be permitted to be located side-by-side.

> **Advisory 221.2.3.1 Horizontal Dispersion.** Horizontal dispersion of wheelchair spaces is the placement of spaces in an assembly facility seating area from side-to-side or, in the case of an arena or stadium, around the field of play or performance area.

221.2.3.2 Vertical Dispersion. *Wheelchair spaces* shall be dispersed vertically at varying distances from the screen, performance area, or playing field. In addition, *wheelchair spaces* shall be located in each balcony or *mezzanine* that is located on an *accessible* route.
EXCEPTIONS: 1. Vertical dispersion shall not be required in *assembly areas* with 300 or fewer seats if the *wheelchair spaces* provide viewing angles that are equivalent to, or better than, the average viewing angle provided in the *facility*.
2. In bleachers, *wheelchair spaces* shall not be required to be provided in rows other than rows at points of entry to bleacher seating.

> **Advisory 221.2.3.2 Vertical Dispersion.** When wheelchair spaces are dispersed vertically in an assembly facility they are placed at different locations within the seating area from front-to-back so that the distance from the screen, stage, playing field, area of sports activity, or other focal point is varied among wheelchair spaces.

> **Advisory 221.2.3.2 Vertical Dispersion Exception 2.** Points of entry to bleacher seating may include, but are not limited to, cross aisles, concourses, vomitories, and entrance ramps and stairs. Vertical, center, or side aisles adjoining bleacher seating that are stepped or tiered are not considered entry points.

221.3 Companion Seats. At least one companion seat complying with 802.3 shall be provided for each *wheelchair space* required by 221.2.1.

221.4 Designated Aisle Seats. At least 5 percent of the total number of aisle seats provided shall comply with 802.4 and shall be the aisle seats located closest to *accessible* routes.
EXCEPTION: Team or player seating areas serving *areas of sport activity* shall not be required to comply with 221.4.

> **Advisory 221.4 Designated Aisle Seats.** When selecting which aisle seats will meet the requirements of 802.4, those aisle seats which are closest to, not necessarily on, accessible routes must be selected first. For example, an assembly area has two aisles (A and B) serving seating areas with an accessible route connecting to the top and bottom of Aisle A only. The aisle seats chosen to meet 802.4 must be those at the top and bottom of Aisle A, working toward the middle. Only when all seats on Aisle A would not meet the five percent minimum would seats on Aisle B be designated.

221.5 Lawn Seating. Lawn seating areas and exterior overflow seating areas, where fixed seats are not provided, shall connect to an *accessible* route.

222 Dressing, Fitting, and Locker Rooms

222.1 General. Where dressing rooms, fitting rooms, or locker rooms are provided, at least 5 percent, but no fewer than one, of each type of use in each cluster provided shall comply with 803.
  EXCEPTION: In *alterations*, where it is *technically infeasible* to provide rooms in accordance with 222.1, one room for each sex on each level shall comply with 803. Where only unisex rooms are provided, unisex rooms shall be permitted.

> **Advisory 222.1 General.** A "cluster" is a group of rooms proximate to one another. Generally, rooms in a cluster are within sight of, or adjacent to, one another. Different styles of design provide users varying levels of privacy and convenience. Some designs include private changing facilities that are close to core areas of the facility, while other designs use space more economically and provide only group dressing facilities. Regardless of the type of facility, dressing, fitting, and locker rooms should provide people with disabilities rooms that are equally private and convenient to those provided others. For example, in a physician's office, if people without disabilities must traverse the full length of the office suite in clothing other than their street clothes, it is acceptable for people with disabilities to be asked to do the same.

222.2 Coat Hooks and Shelves. Where coat hooks or shelves are provided in dressing, fitting or locker rooms without individual compartments, at least one of each type shall comply with 803.5. Where coat hooks or shelves are provided in individual compartments at least one of each type complying with 803.5 shall be provided in individual compartments in dressing, fitting, or locker rooms required to comply with 222.1.

223 Medical Care and Long-Term Care Facilities

223.1 General. In licensed medical care *facilities* and licensed long-term care *facilities* where the period of stay exceeds twenty-four hours, patient or resident sleeping rooms shall be provided in accordance with 223. **[See additional requirements at 28 CFR 35.151(h), p. 13, and 28 CFR 36.406(g), p. 30.]**
  EXCEPTION: Toilet rooms that are part of critical or intensive care patient sleeping rooms shall not be required to comply with 603.

> **Advisory 223.1 General.** Because medical facilities frequently reconfigure spaces to reflect changes in medical specialties, Section 223.1 does not include a provision for dispersion of accessible patient or resident sleeping rooms. The lack of a design requirement does not mean that covered entities are not required to provide services to people with disabilities where accessible rooms are not dispersed in specialty areas. Locate accessible rooms near core areas that are less likely to change over time. While dispersion is not required, the flexibility it provides can be a critical factor in ensuring cost effective compliance with applicable civil rights laws, including titles II and III of the ADA and Section 504 of the Rehabilitation Act of 1973, as amended.

> **Advisory 223.1 General (Continued).** Additionally, all types of features and amenities should be dispersed among accessible sleeping rooms to ensure equal access to and a variety of choices for all patients and residents.

223.1.1 Alterations. Where sleeping rooms are *altered* or *added*, the requirements of 223 shall apply only to the sleeping rooms being *altered* or *added* until the number of sleeping rooms complies with the minimum number required for new construction.

> **Advisory 223.1.1 Alterations.** In alterations and additions, the minimum required number is based on the total number of sleeping rooms altered or added instead of on the total number of sleeping rooms provided in a facility. As a facility is altered over time, every effort should be made to disperse accessible sleeping rooms among patient care areas such as pediatrics, cardiac care, maternity, and other units. In this way, people with disabilities can have access to the full-range of services provided by a medical care facility.

223.2 Hospitals, Rehabilitation Facilities, Psychiatric Facilities and Detoxification Facilities. Hospitals, rehabilitation *facilities*, psychiatric *facilities* and detoxification *facilities* shall comply with 223.2.

223.2.1 Facilities Not Specializing in Treating Conditions That Affect Mobility. In *facilities* not specializing in treating conditions that affect mobility, at least 10 percent, but no fewer than one, of the patient sleeping rooms shall provide mobility features complying with 805.

223.2.2 Facilities Specializing in Treating Conditions That Affect Mobility. In *facilities* specializing in treating conditions that affect mobility, 100 percent of the patient sleeping rooms shall provide mobility features complying with 805.

> **Advisory 223.2.2 Facilities Specializing in Treating Conditions That Affect Mobility.** Conditions that affect mobility include conditions requiring the use or assistance of a brace, cane, crutch, prosthetic device, wheelchair, or powered mobility aid; arthritic, neurological, or orthopedic conditions that severely limit one's ability to walk; respiratory diseases and other conditions which may require the use of portable oxygen; and cardiac conditions that impose significant functional limitations. Facilities that may provide treatment for, but that do not specialize in treatment of such conditions, such as general rehabilitation hospitals, are not subject to this requirement but are subject to Section 223.2.1.

223.3 Long-Term Care Facilities. In licensed long-term care *facilities*, at least 50 percent, but no fewer than one, of each type of resident sleeping room shall provide mobility features complying with 805.

224 Transient Lodging Guest Rooms

224.1 General. *Transient lodging facilities* shall provide guest rooms in accordance with 224.
   [See additional requirements for places of lodging at 28 CFR 36.406(c), p. 28. and for housing at a place of education at 28 CFR 35.151(f), p. 11, and 28 CFR 36.406(e), p. 29.]

Advisory 224.1 General. Certain facilities used for transient lodging, including time shares, dormitories, and town homes may be covered by both these requirements and the Fair Housing Amendments Act. The Fair Housing Amendments Act requires that certain residential structures having four or more multi-family dwelling units, regardless of whether they are privately owned or federally assisted, include certain features of accessible and adaptable design according to guidelines established by the U.S. Department of Housing and Urban Development (HUD). This law and the appropriate regulations should be consulted before proceeding with the design and construction of residential housing.

224.1.1 Alterations. Where guest rooms are *altered* or *added*, the requirements of 224 shall apply only to the guest rooms being *altered* or *added* until the number of guest rooms complies with the minimum number required for new construction.

Advisory 224.1.1 Alterations. In alterations and additions, the minimum required number of accessible guest rooms is based on the total number of guest rooms altered or added instead of the total number of guest rooms provided in a facility. Typically, each alteration of a facility is limited to a particular portion of the facility. When accessible guest rooms are added as a result of subsequent alterations, compliance with 224.5 (Dispersion) is more likely to be achieved if all of the accessible guest rooms are not provided in the same area of the facility.

224.1.2 Guest Room Doors and Doorways. *Entrances,* doors, and doorways providing user passage into and within guest rooms that are not required to provide mobility features complying with 806.2 shall comply with 404.2.3.

    EXCEPTION: Shower and sauna doors in guest rooms that are not required to provide mobility features complying with 806.2 shall not be required to comply with 404.2.3.

Advisory 224.1.2 Guest Room Doors and Doorways. Because of the social interaction that often occurs in lodging facilities, an accessible clear opening width is required for doors and doorways to and within all guest rooms, including those not required to be accessible. This applies to all doors, including bathroom doors, that allow full user passage. Other requirements for doors and doorways in Section 404 do not apply to guest rooms not required to provide mobility features.

224.2 Guest Rooms with Mobility Features. In *transient lodging facilities*, guest rooms with mobility features complying with 806.2 shall be provided in accordance with Table 224.2.

Table 224.2 Guest Rooms with Mobility Features

Total Number of Guest Rooms Provided	Minimum Number of Required Rooms Without Roll-in Showers	Minimum Number of Required Rooms With Roll-in Showers	Total Number of Required Rooms
1 to 25	1	0	1
26 to 50	2	0	2
51 to 75	3	1	4
76 to 100	4	1	5
101 to 150	5	2	7
151 to 200	6	2	8
201 to 300	7	3	10
301 to 400	8	4	12
401 to 500	9	4	13
501 to 1000	2 percent of total	1 percent of total	3 percent of total
1001 and over	20, plus 1 for each 100, or fraction thereof, over 1000	10, plus 1 for each 100, or fraction thereof, over 1000	30, plus 2 for each 100, or fraction thereof, over 1000

224.3 Beds. In guest rooms having more than 25 beds, 5 percent minimum of the beds shall have clear floor *space* complying with 806.2.3.

224.4 Guest Rooms with Communication Features. In *transient lodging facilities*, guest rooms with communication features complying with 806.3 shall be provided in accordance with Table 224.4.

Table 224.4 Guest Rooms with Communication Features

Total Number of Guest Rooms Provided	Minimum Number of Required Guest Rooms With Communication Features
2 to 25	2
26 to 50	4
51 to 75	7
76 to 100	9
101 to 150	12

Table 224.4 Guest Rooms with Communication Features

Total Number of Guest Rooms Provided	Minimum Number of Required Guest Rooms With Communication Features
151 to 200	14
201 to 300	17
301 to 400	20
401 to 500	22
501 to 1000	5 percent of total
1001 and over	50, plus 3 for each 100 over 1000

224.5 Dispersion. Guest rooms required to provide mobility features complying with 806.2 and guest rooms required to provide communication features complying with 806.3 shall be dispersed among the various classes of guest rooms, and shall provide choices of types of guest rooms, number of beds, and other amenities comparable to the choices provided to other guests. Where the minimum number of guest rooms required to comply with 806 is not sufficient to allow for complete dispersion, guest rooms shall be dispersed in the following priority: guest room type, number of beds, and amenities. At least one guest room required to provide mobility features complying with 806.2 shall also provide communication features complying with 806.3. Not more than 10 percent of guest rooms required to provide mobility features complying with 806.2 shall be used to satisfy the minimum number of guest rooms required to provide communication features complying with 806.3.

> **Advisory 224.5 Dispersion.** Factors to be considered in providing an equivalent range of options may include, but are not limited to, room size, bed size, cost, view, bathroom fixtures such as hot tubs and spas, smoking and nonsmoking, and the number of rooms provided.

225 Storage

225.1 General. Storage *facilities* shall comply with 225.

225.2 Storage. Where storage is provided in accessible *spaces*, at least one of each type shall comply with 811.

> **Advisory 225.2 Storage.** Types of storage include, but are not limited to, closets, cabinets, shelves, clothes rods, hooks, and drawers. Where provided, at least one of each type of storage must be within the reach ranges specified in 308; however, it is permissible to install additional storage outside the reach ranges.

225.2.1 Lockers. Where lockers are provided, at least 5 percent, but no fewer than one of each type, shall comply with 811.

> **Advisory 225.2.1 Lockers.** Different types of lockers may include full-size and half-size lockers, as well as those specifically designed for storage of various sports equipment.

225.2.2 Self-Service Shelving. Self-service shelves shall be located on an *accessible* route complying with 402. Self-service shelving shall not be required to comply with 308.

> **Advisory 225.2.2 Self-Service Shelving.** Self-service shelves include, but are not limited to, library, store, or post office shelves.

225.3 Self-Service Storage Facilities. *Self-service storage facilities* shall provide individual *self-service storage spaces* complying with these requirements in accordance with Table 225.3.

Table 225.3 Self-Service Storage Facilities

Total Spaces in Facility	Minimum Number of Spaces Required to be Accessible
1 to 200	5 percent, but no fewer than 1
201 and over	10, plus 2 percent of total number of units over 200

> **Advisory 225.3 Self-Service Storage Facilities.** Although there are no technical requirements that are unique to self-service storage facilities, elements and spaces provided in facilities containing self-service storage spaces required to comply with these requirements must comply with this document where applicable. For example: the number of storage spaces required to comply with these requirements must provide Accessible Routes complying with Section 206; Accessible Means of Egress complying with Section 207; Parking Spaces complying with Section 208; and, where provided, other public use or common use elements and facilities such as toilet rooms, drinking fountains, and telephones must comply with the applicable requirements of this document.

225.3.1 Dispersion. Individual *self-service storage spaces* shall be dispersed throughout the various classes of *spaces* provided. Where more classes of *spaces* are provided than the number required to be *accessible*, the number of *spaces* shall not be required to exceed that required by Table 225.3. *Self-service storage spaces* complying with Table 225.3 shall not be required to be dispersed among *buildings* in a multi-*building facility*.

226 Dining Surfaces and Work Surfaces

226.1 General. Where dining surfaces are provided for the consumption of food or drink, at least 5 percent of the seating *spaces* and standing *spaces* at the dining surfaces shall comply with 902. In addition, where work surfaces are provided for use by other than employees, at least 5 percent shall comply with 902.
    EXCEPTIONS: 1. Sales counters and service counters shall not be required to comply with 902.

2. Check writing surfaces provided at check-out aisles not required to comply with 904.3 shall not be required to comply with 902.

> **Advisory 226.1 General.** In facilities covered by the ADA, this requirement does not apply to work surfaces used only by employees. However, the ADA and, where applicable, Section 504 of the Rehabilitation Act of 1973, as amended, provide that employees are entitled to "reasonable accommodations." With respect to work surfaces, this means that employers may need to procure or adjust work stations such as desks, laboratory and work benches, fume hoods, reception counters, teller windows, study carrels, commercial kitchen counters, and conference tables to accommodate the individual needs of employees with disabilities on an "as needed" basis. Consider work surfaces that are flexible and permit installation at variable heights and clearances.

226.2 Dispersion. Dining surfaces and work surfaces required to comply with 902 shall be dispersed throughout the *space* or *facility* containing dining surfaces and work surfaces.

227 Sales and Service

227.1 General. Where provided, check-out aisles, sales counters, service counters, food service lines, queues, and waiting lines shall comply with 227 and 904.

227.2 Check-Out Aisles. Where check-out aisles are provided, check-out aisles complying with 904.3 shall be provided in accordance with Table 227.2. Where check-out aisles serve different functions, check-out aisles complying with 904.3 shall be provided in accordance with Table 227.2 for each function. Where check-out aisles are dispersed throughout the *building* or *facility*, check-out aisles complying with 904.3 shall be dispersed.

    EXCEPTION: Where the selling *space* is under 5000 square feet (465 m^2) no more than one check-out aisle complying with 904.3 shall be required.

Table 227.2 Check-Out Aisles

Number of Check-Out Aisles of Each Function	Minimum Number of Check-Out Aisles of Each Function Required to Comply with 904.3
1 to 4	1
5 to 8	2
9 to 15	3
16 and over	3, plus 20 percent of additional aisles

227.2.1 Altered Check-Out Aisles. Where check-out aisles are *altered*, at least one of each check-out aisle serving each function shall comply with 904.3 until the number of check-out aisles complies with 227.2.

227.3 Counters. Where provided, at least one of each type of sales counter and service counter shall comply with 904.4. Where counters are dispersed throughout the *building* or *facility*, counters complying with 904.4 also shall be dispersed.

> **Advisory 227.3 Counters.** Types of counters that provide different services in the same facility include, but are not limited to, order, pick-up, express, and returns. One continuous counter can be used to provide different types of service. For example, order and pick-up are different services. It would not be acceptable to provide access only to the part of the counter where orders are taken when orders are picked-up at a different location on the same counter. Both the order and pick-up section of the counter must be accessible.

227.4 Food Service Lines. Food service lines shall comply with 904.5. Where self-service shelves are provided, at least 50 percent, but no fewer than one, of each type provided shall comply with 308.

227.5 Queues and Waiting Lines. Queues and waiting lines servicing counters or check-out aisles required to comply with 904.3 or 904.4 shall comply with 403.

228 Depositories, Vending Machines, Change Machines, Mail Boxes, and Fuel Dispensers

228.1 General. Where provided, at least one of each type of depository, vending machine, change machine, and fuel dispenser shall comply with 309.

EXCEPTION: Drive-up only depositories shall not be required to comply with 309.

> **Advisory 228.1 General.** Depositories include, but are not limited to, night receptacles in banks, post offices, video stores, and libraries.

228.2 Mail Boxes. Where *mail boxes* are provided in an interior location, at least 5 percent, but no fewer than one, of each type shall comply with 309. In residential *facilities*, where *mail boxes* are provided for each *residential dwelling unit*, *mail boxes* complying with 309 shall be provided for each *residential dwelling unit* required to provide mobility features complying with 809.2 through 809.4.

229 Windows

229.1 General. Where glazed openings are provided in *accessible* rooms or *spaces* for operation by occupants, at least one opening shall comply with 309. Each glazed opening required by an *administrative authority* to be operable shall comply with 309.

EXCEPTION: 1. Glazed openings in *residential dwelling units* required to comply with 809 shall not be required to comply with 229.

2. Glazed openings in guest rooms required to provide communication features and in guest rooms required to comply with 206.5.3 shall not be required to comply with 229.

230 Two-Way Communication Systems

230.1 General. Where a two-way communication system is provided to gain admittance to a *building* or *facility* or to restricted areas within a *building* or *facility*, the system shall comply with 708.

> **Advisory 230.1 General.** This requirement applies to facilities such as office buildings, courthouses, and other facilities where admittance to the building or restricted spaces is dependent on two-way communication systems.

231 Judicial Facilities

231.1 General. Judicial *facilities* shall comply with 231.

231.2 Courtrooms. Each courtroom shall comply with 808.

231.3 Holding Cells. Where provided, central holding cells and court-floor holding cells shall comply with 231.3.

>**231.3.1 Central Holding Cells.** Where separate central holding cells are provided for adult male, juvenile male, adult female, or juvenile female, one of each type shall comply with 807.2. Where central holding cells are provided and are not separated by age or sex, at least one cell complying with 807.2 shall be provided.
>
>**231.3.2 Court-Floor Holding Cells.** Where separate court-floor holding cells are provided for adult male, juvenile male, adult female, or juvenile female, each courtroom shall be served by one cell of each type complying with 807.2. Where court-floor holding cells are provided and are not separated by age or sex, courtrooms shall be served by at least one cell complying with 807.2. Cells may serve more than one courtroom.

231.4 Visiting Areas. Visiting areas shall comply with 231.4.

>**231.4.1 Cubicles and Counters.** At least 5 percent, but no fewer than one, of cubicles shall comply with 902 on both the visitor and detainee sides. Where counters are provided, at least one shall comply with 904.4.2 on both the visitor and detainee sides.
>   **EXCEPTION:** The detainee side of cubicles or counters at non-contact visiting areas not serving holding cells required to comply with 231 shall not be required to comply with 902 or 904.4.2.
>
>**231.4.2 Partitions.** Where solid partitions or security glazing separate visitors from detainees at least one of each type of cubicle or counter partition shall comply with 904.6.

232 Detention Facilities and Correctional Facilities

232.1 General. *Buildings, facilities*, or portions thereof, in which people are detained for penal or correction purposes, or in which the liberty of the inmates is restricted for security reasons shall comply with 232. **[See additional requirements at 28 CFR 35.151(k), p. 13.]**

> **Advisory 232.1 General.** Detention facilities include, but are not limited to, jails, detention centers, and holding cells in police stations. Correctional facilities include, but are not limited to, prisons, reformatories, and correctional centers.

232.2 General Holding Cells and General Housing Cells. General holding cells and general housing cells shall be provided in accordance with 232.2.

> **EXCEPTION:** *Alterations* to cells shall not be required to comply except to the extent determined by the Attorney General.

> **Advisory 232.2 General Holding Cells and General Housing Cells.** Accessible cells or rooms should be dispersed among different levels of security, housing categories, and holding classifications (e.g., male/female and adult/juvenile) to facilitate access. Many detention and correctional facilities are designed so that certain areas (e.g., "shift" areas) can be adapted to serve as different types of housing according to need. For example, a shift area serving as a medium-security housing unit might be redesignated for a period of time as a high-security housing unit to meet capacity needs. Placement of accessible cells or rooms in shift areas may allow additional flexibility in meeting requirements for dispersion of accessible cells or rooms.
>
> **Advisory 232.2 General Holding Cells and General Housing Cells Exception.** Although these requirements do not specify that cells be accessible as a consequence of an alteration, title II of the ADA requires that each service, program, or activity conducted by a public entity, when viewed in its entirety, be readily accessible to and usable by individuals with disabilities. This requirement must be met unless doing so would fundamentally alter the nature of a service, program, or activity or would result in undue financial and administrative burdens.

232.2.1 Cells with Mobility Features. At least 2 percent, but no fewer than one, of the total number of cells in a *facility* shall provide mobility features complying with 807.2.

232.2.1.1 Beds. In cells having more than 25 beds, at least 5 percent of the beds shall have clear floor *space* complying with 807.2.3.

232.2.2 Cells with Communication Features. At least 2 percent, but no fewer than one, of the total number of general holding cells and general housing cells equipped with audible emergency alarm systems and permanently installed telephones within the cell shall provide communication features complying with 807.3.

232.3 Special Holding Cells and Special Housing Cells. Where special holding cells or special housing cells are provided, at least one cell serving each purpose shall provide mobility features complying with 807.2. Cells subject to this requirement include, but are not limited to, those used for purposes of orientation, protective custody, administrative or disciplinary detention or segregation, detoxification, and medical isolation.

> **EXCEPTION:** *Alterations* to cells shall not be required to comply except to the extent determined by the Attorney General.

232.4 Medical Care Facilities. Patient bedrooms or cells required to comply with 223 shall be provided in addition to any medical isolation cells required to comply with 232.3.

232.5 Visiting Areas. Visiting areas shall comply with 232.5.

232.5.1 Cubicles and Counters. At least 5 percent, but no fewer than one, of cubicles shall comply with 902 on both the visitor and detainee sides. Where counters are provided, at least one shall comply with 904.4.2 on both the visitor and detainee or inmate sides.
  EXCEPTION: The inmate or detainee side of cubicles or counters at non-contact visiting areas not serving holding cells or housing cells required to comply with 232 shall not be required to comply with 902 or 904.4.2.

232.5.2 Partitions. Where solid partitions or security glazing separate visitors from detainees or inmates at least one of each type of cubicle or counter partition shall comply with 904.6.

233 Residential Facilities

233.1 General. *Facilities* with *residential dwelling units* shall comply with 233. [See additional requirements at 28 CFR 35.151(e) and (f), p. 11, and 28 CFR 36.406(d) and (e), pp. 28 and 29.]

> **Advisory 233.1 General.** Section 233 outlines the requirements for residential facilities subject to the Americans with Disabilities Act of 1990. The facilities covered by Section 233, as well as other facilities not covered by this section, may still be subject to other Federal laws such as the Fair Housing Act and Section 504 of the Rehabilitation Act of 1973, as amended. For example, the Fair Housing Act requires that certain residential structures having four or more multi-family dwelling units, regardless of whether they are privately owned or federally assisted, include certain features of accessible and adaptable design according to guidelines established by the U.S. Department of Housing and Urban Development (HUD). These laws and the appropriate regulations should be consulted before proceeding with the design and construction of residential facilities.
>
> Residential facilities containing residential dwelling units provided by entities subject to HUD's Section 504 regulations and residential dwelling units covered by Section 233.3 must comply with the technical and scoping requirements in Chapters 1 through 10 included this document. Section 233 is not a stand-alone section; this section only addresses the minimum number of residential dwelling units within a facility required to comply with Chapter 8. However, residential facilities must also comply with the requirements of this document. For example: Section 206.5.4 requires all doors and doorways providing user passage in residential dwelling units providing mobility features to comply with Section 404; Section 206.7.6 permits platform lifts to be used to connect levels within residential dwelling units providing mobility features; Section 208 provides general scoping for accessible parking and Section 208.2.3.1 specifies the required number of accessible parking spaces for each residential dwelling unit providing mobility features; Section 228.2 requires mail boxes to be within reach ranges when they serve residential dwelling units providing mobility features; play areas are addressed in Section 240; and swimming pools are addressed in Section 242. There are special provisions applicable to facilities containing residential dwelling units at: Exception 3 to 202.3; Exception to 202.4; 203.8; and Exception 4 to 206.2.3.

233.2 Residential Dwelling Units Provided by Entities Subject to HUD Section 504 Regulations. Where *facilities* with *residential dwelling units* are provided by entities subject to regulations issued by the Department of Housing and Urban Development (HUD) under Section 504 of the Rehabilitation Act of 1973, as amended, such entities shall provide *residential dwelling units* with mobility features complying with 809.2 through 809.4 in a number required by the applicable HUD regulations. *Residential dwelling units* required to provide mobility features complying with 809.2 through 809.4 shall be on an *accessible* route as required by 206. In addition, such entities shall provide *residential dwelling units* with communication features complying with 809.5 in a number required by the applicable HUD regulations. Entities subject to 233.2 shall not be required to comply with 233.3.

> **Advisory 233.2 Residential Dwelling Units Provided by Entities Subject to HUD Section 504 Regulations.** Section 233.2 requires that entities subject to HUD's regulations implementing Section 504 of the Rehabilitation Act of 1973, as amended, provide residential dwelling units containing mobility features and residential dwelling units containing communication features complying with these regulations in a number specified in HUD's Section 504 regulations. Further, the residential dwelling units provided must be dispersed according to HUD's Section 504 criteria. In addition, Section 233.2 defers to HUD the specification of criteria by which the technical requirements of this document will apply to alterations of existing facilities subject to HUD's Section 504 regulations.

233.3 Residential Dwelling Units Provided by Entities Not Subject to HUD Section 504 Regulations. *Facilities* with *residential dwelling units* provided by entities not subject to regulations issued by the Department of Housing and Urban Development (HUD) under Section 504 of the Rehabilitation Act of 1973, as amended, shall comply with 233.3.

233.3.1 Minimum Number: New Construction. Newly constructed *facilities* with *residential dwelling units* shall comply with 233.3.1.

EXCEPTION: Where *facilities* contain 15 or fewer *residential dwelling units*, the requirements of 233.3.1.1 and 233.3.1.2 shall apply to the total number of *residential dwelling units* that are constructed under a single contract, or are developed as a whole, whether or not located on a common *site*.

233.3.1.1 Residential Dwelling Units with Mobility Features. In *facilities* with *residential dwelling units*, at least 5 percent, but no fewer than one unit, of the total number of *residential dwelling units* shall provide mobility features complying with 809.2 through 809.4 and shall be on an *accessible* route as required by 206.

233.3.1.2 Residential Dwelling Units with Communication Features. In *facilities* with *residential dwelling units*, at least 2 percent, but no fewer than one unit, of the total number of *residential dwelling units* shall provide communication features complying with 809.5.

233.3.2 Residential Dwelling Units for Sale. *Residential dwelling units* offered for sale shall provide *accessible* features to the extent required by regulations issued by Federal agencies under the Americans with Disabilities Act or Section 504 of the Rehabilitation Act of 1973, as amended. **[See additional requirements at 28 CFR 35.151(j), p. 13.]**

> **Advisory 233.3.2 Residential Dwelling Units for Sale.** A public entity that conducts a program to build housing for purchase by individual home buyers must provide access according to the requirements of the ADA regulations and a program receiving Federal financial assistance must comply with the applicable Section 504 regulation.

233.3.3 Additions. Where an *addition* to an existing *building* results in an increase in the number of *residential dwelling units*, the requirements of 233.3.1 shall apply only to the *residential dwelling units* that are *added* until the total number of *residential dwelling units* complies with the minimum number required by 233.3.1. *Residential dwelling units* required to comply with 233.3.1.1 shall be on an *accessible* route as required by 206.

233.3.4 Alterations. *Alterations* shall comply with 233.3.4.
  EXCEPTION: Where compliance with 809.2, 809.3, or 809.4 is *technically infeasible*, or where it is *technically infeasible* to provide an *accessible* route to a *residential dwelling unit*, the entity shall be permitted to *alter* or construct a comparable *residential dwelling unit* to comply with 809.2 through 809.4 provided that the minimum number of *residential dwelling units* required by 233.3.1.1 and 233.3.1.2, as applicable, is satisfied.

> **Advisory 233.3.4 Alterations Exception.** A substituted dwelling unit must be comparable to the dwelling unit that is not made accessible. Factors to be considered in comparing one dwelling unit to another should include the number of bedrooms; amenities provided within the dwelling unit; types of common spaces provided within the facility; and location with respect to community resources and services, such as public transportation and civic, recreational, and mercantile facilities.

233.3.4.1 Alterations to Vacated Buildings. Where a *building* is vacated for the purposes of *alteration*, and the *altered building* contains more than 15 *residential dwelling units*, at least 5 percent of the *residential dwelling units* shall comply with 809.2 through 809.4 and shall be on an *accessible* route as required by 206. In addition, at least 2 percent of the *residential dwelling units* shall comply with 809.5.

> **Advisory 233.3.4.1 Alterations to Vacated Buildings.** This provision is intended to apply where a building is vacated with the intent to alter the building. Buildings that are vacated solely for pest control or asbestos removal are not subject to the requirements to provide residential dwelling units with mobility features or communication features.

233.3.4.2 Alterations to Individual Residential Dwelling Units. In individual *residential dwelling units*, where a bathroom or a kitchen is substantially *altered*, and at least one other room is *altered*, the requirements of 233.3.1 shall apply to the *altered residential dwelling units* until the total number of *residential dwelling units* complies with the minimum number required by 233.3.1.1 and 233.3.1.2. *Residential dwelling units* required to comply with 233.3.1.1 shall be on an *accessible* route as required by 206.
  EXCEPTION: Where *facilities* contain 15 or fewer *residential dwelling units*, the requirements of 233.3.1.1 and 233.3.1.2 shall apply to the total number of *residential dwelling units* that are *altered* under a single contract, or are developed as a whole, whether or not located on a common *site*.

> **Advisory 233.3.4.2 Alterations to Individual Residential Dwelling Units.** Section 233.3.4.2 uses the terms "substantially altered" and "altered." A substantial alteration to a kitchen or bathroom includes, but is not limited to, alterations that are changes to or rearrangements in the plan configuration, or replacement of cabinetry. Substantial alterations do not include normal maintenance or appliance and fixture replacement, unless such maintenance or replacement requires changes to or rearrangements in the plan configuration, or replacement of cabinetry. The term "alteration" is defined both in Section 106 of these requirements and in the Department of Justice ADA regulations.

233.3.5 Dispersion. *Residential dwelling units* required to provide mobility features complying with 809.2 through 809.4 and *residential dwelling units* required to provide communication features complying with 809.5 shall be dispersed among the various types of *residential dwelling units* in the *facility* and shall provide choices of *residential dwelling units* comparable to, and integrated with, those available to other residents.

EXCEPTION: Where multi-*story residential dwelling units* are one of the types of *residential dwelling units* provided, one-*story residential dwelling units* shall be permitted as a substitute for multi-*story residential dwelling units* where equivalent *spaces* and amenities are provided in the one-*story residential dwelling unit*.

234 Amusement Rides

234.1 General. A*musement rides* shall comply with 234.

EXCEPTION: Mobile or portable *amusement rides* shall not be required to comply with 234.

> **Advisory 234.1 General.** These requirements apply generally to newly designed and constructed amusement rides and attractions. A custom designed and constructed ride is new upon its first use, which is the first time amusement park patrons take the ride. With respect to amusement rides purchased from other entities, new refers to the first permanent installation of the ride, whether it is used off the shelf or modified before it is installed. Where amusement rides are moved after several seasons to another area of the park or to another park, the ride would not be considered newly designed or newly constructed.
>
> Some amusement rides and attractions that have unique designs and features are not addressed by these requirements. In those situations, these requirements are to be applied to the extent possible. An example of an amusement ride not specifically addressed by these requirements includes "virtual reality" rides where the device does not move through a fixed course within a defined area. An accessible route must be provided to these rides. Where an attraction or ride has unique features for which there are no applicable scoping provisions, then a reasonable number, but at least one, of the features must be located on an accessible route. Where there are appropriate technical provisions, they must be applied to the elements that are covered by the scoping provisions.
>
> **Advisory 234.1 General Exception.** Mobile or temporary rides are those set up for short periods of time such as traveling carnivals, State and county fairs, and festivals. The amusement rides that are covered by 234.1 are ones that are not regularly assembled and disassembled.

234.2 Load and Unload Areas. Load and unload areas serving *amusement rides* shall comply with 1002.3.

234.3 Minimum Number. *Amusement rides* shall provide at least one *wheelchair space* complying with 1002.4, or at least one *amusement ride seat* designed for transfer complying with 1002.5, or at least one *transfer device* complying with 1002.6.
   EXCEPTIONS: 1. *Amusement rides* that are controlled or operated by the rider shall not be required to comply with 234.3.
   2. *Amusement rides* designed primarily for children, where children are assisted on and off the ride by an adult, shall not be required to comply with 234.3.
   3. *Amusement rides* that do not provide *amusement ride seats* shall not be required to comply with 234.3.

> Advisory 234.3 Minimum Number Exceptions 1 through 3. Amusement rides controlled or operated by the rider, designed for children, or rides without ride seats are not required to comply with 234.3. These rides are not exempt from the other provisions in 234 requiring an accessible route to the load and unload areas and to the ride. The exception does not apply to those rides where patrons may cause the ride to make incidental movements, but where the patron otherwise has no control over the ride.
>
> Advisory 234.3 Minimum Number Exception 2. The exception is limited to those rides designed "primarily" for children, where children are assisted on and off the ride by an adult. This exception is limited to those rides designed for children and not for the occasional adult user. An accessible route to and turning space in the load and unload area will provide access for adults and family members assisting children on and off these rides.

234.4 Existing Amusement Rides. Where existing *amusement rides* are *altered*, the *alteration* shall comply with 234.4.

> Advisory 234.4 Existing Amusement Rides. Routine maintenance, painting, and changing of theme boards are examples of activities that do not constitute an alteration subject to this section.

234.4.1 Load and Unload Areas. Where load and unload areas serving existing *amusement rides* are newly designed and constructed, the load and unload areas shall comply with 1002.3.

234.4.2 Minimum Number. Where the structural or operational characteristics of an *amusement ride* are *altered* to the extent that the *amusement ride*'s performance differs from that specified by the manufacturer or the original design, the *amusement ride* shall comply with 234.3.

235 Recreational Boating Facilities

235.1 General. Recreational boating *facilities* shall comply with 235.

235.2 Boat Slips. *Boat slips* complying with 1003.3.1 shall be provided in accordance with Table 235.2. Where the number of *boat slips* is not identified, each 40 feet (12 m) of *boat slip* edge provided along the perimeter of the pier shall be counted as one *boat slip* for the purpose of this section.

Table 235.2 Boat Slips

Total Number of Boat Slips Provided in Facility	Minimum Number of Required Accessible Boat Slips
1 to 25	1
26 to 50	2
51 to 100	3
101 to 150	4
151 to 300	5
301 to 400	6
401 to 500	7
501 to 600	8
601 to 700	9
701 to 800	10
801 to 900	11
901 to 1000	12
1001 and over	12, plus 1 for every 100, or fraction thereof, over 1000

Advisory 235.2 Boat Slips. The requirement for boat slips also applies to piers where boat slips are not demarcated. For example, a single pier 25 feet (7620 mm) long and 5 feet (1525 mm) wide (the minimum width specified by Section 1003.3) allows boats to moor on three sides. Because the number of boat slips is not demarcated, the total length of boat slip edge (55 feet, 17 m) must be used to determine the number of boat slips provided (two). This number is based on the specification in Section 235.2 that each 40 feet (12 m) of boat slip edge, or fraction thereof, counts as one boat slip. In this example, Table 235.2 would require one boat slip to be accessible.

235.2.1 Dispersion. *Boat slips* complying with 1003.3.1 shall be dispersed throughout the various types of *boat slips* provided. Where the minimum number of *boat slips* required to comply with 1003.3.1 has been met, no further dispersion shall be required.

> **Advisory 235.2.1 Dispersion.** Types of boat slips are based on the size of the boat slips; whether single berths or double berths, shallow water or deep water, transient or longer-term lease, covered or uncovered; and whether slips are equipped with features such as telephone, water, electricity or cable connections. The term "boat slip" is intended to cover any pier area other than launch ramp boarding piers where recreational boats are moored for purposes of berthing, embarking, or disembarking. For example, a fuel pier may contain boat slips, and this type of short term slip would be included in determining compliance with 235.2.

235.3 Boarding Piers at Boat Launch Ramps. Where *boarding piers* are provided at *boat launch ramps*, at least 5 percent, but no fewer than one, of the *boarding piers* shall comply with 1003.3.2.

236 Exercise Machines and Equipment

236.1 General. At least one of each type of exercise machine and equipment shall comply with 1004.

> **Advisory 236.1 General.** Most strength training equipment and machines are considered different types. Where operators provide a biceps curl machine and cable-cross-over machine, both machines are required to meet the provisions in this section, even though an individual may be able to work on their biceps through both types of equipment.
>
> Similarly, there are many types of cardiovascular exercise machines, such as stationary bicycles, rowing machines, stair climbers, and treadmills. Each machine provides a cardiovascular exercise and is considered a different type for purposes of these requirements.

237 Fishing Piers and Platforms

237.1 General. Fishing piers and platforms shall comply with 1005.

238 Golf Facilities

238.1 General. Golf *facilities* shall comply with 238.

238.2 Golf Courses. Golf courses shall comply with 238.2.

238.2.1 Teeing Grounds. Where one *teeing ground* is provided for a hole, the *teeing ground* shall be designed and constructed so that a golf car can enter and exit the *teeing ground*. Where two *teeing grounds* are provided for a hole, the forward *teeing ground* shall be designed and constructed so that a golf car can enter and exit the *teeing ground*. Where three or more *teeing grounds* are provided for a hole, at least two *teeing grounds*, including the forward *teeing ground*, shall be designed and constructed so that a golf car can enter and exit each *teeing ground*.

> **EXCEPTION:** In existing golf courses, the forward *teeing ground* shall not be required to be one of the *teeing grounds* on a hole designed and constructed so that a golf car can enter and exit the *teeing ground* where compliance is not feasible due to terrain.

238.2.2 Putting Greens. Putting greens shall be designed and constructed so that a golf car can enter and exit the putting green.

238.2.3 Weather Shelters. Where provided, weather shelters shall be designed and constructed so that a golf car can enter and exit the weather shelter and shall comply with 1006.4.

238.3 Practice Putting Greens, Practice Teeing Grounds, and Teeing Stations at Driving Ranges. At least 5 percent, but no fewer than one, of practice putting greens, practice *teeing grounds*, and teeing stations at driving ranges shall be designed and constructed so that a golf car can enter and exit the practice putting greens, practice *teeing grounds*, and teeing stations at driving ranges.

239 Miniature Golf Facilities

239.1 General. Miniature golf *facilities* shall comply with 239.

239.2 Minimum Number. At least 50 percent of holes on miniature golf courses shall comply with 1007.3.

> **Advisory 239.2 Minimum Number.** Where possible, providing access to all holes on a miniature golf course is recommended. If a course is designed with the minimum 50 percent accessible holes, designers or operators are encouraged to select holes which provide for an equivalent experience to the maximum extent possible.

239.3 Miniature Golf Course Configuration. Miniature golf courses shall be configured so that the holes complying with 1007.3 are consecutive. Miniature golf courses shall provide an *accessible* route from the last hole complying with 1007.3 to the course *entrance* or exit without requiring travel through any other holes on the course.
    EXCEPTION: One break in the sequence of consecutive holes shall be permitted provided that the last hole on the miniature golf course is the last hole in the sequence.

> **Advisory 239.3 Miniature Golf Course Configuration.** Where only the minimum 50 percent of the holes are accessible, an accessible route from the last accessible hole to the course exit or entrance must not require travel back through other holes. In some cases, this may require an additional accessible route. Other options include increasing the number of accessible holes in a way that limits the distance needed to connect the last accessible hole with the course exit or entrance.

240 Play Areas

240.1 General. *Play areas* for children ages 2 and over shall comply with 240. Where separate *play areas* are provided within a *site* for specific age groups, each *play area* shall comply with 240.
    EXCEPTIONS: 1. *Play areas* located in family child care *facilities* where the proprietor actually resides shall not be required to comply with 240.
    2. In existing *play areas*, where *play components* are relocated for the purposes of creating safe *use zones* and the ground surface is not *altered* or extended for more than one *use zone*, the *play area* shall not be required to comply with 240.

3. *Amusement attractions* shall not be required to comply with 240.
4. Where *play components* are *altered* and the ground surface is not *altered*, the ground surface shall not be required to comply with 1008.2.6 unless required by 202.4.

> **Advisory 240.1 General.** Play areas may be located on exterior sites or within a building. Where separate play areas are provided within a site for children in specified age groups (e.g., preschool (ages 2 to 5) and school age (ages 5 to 12)), each play area must comply with this section. Where play areas are provided for the same age group on a site but are geographically separated (e.g., one is located next to a picnic area and another is located next to a softball field), they are considered separate play areas and each play area must comply with this section.

240.1.1 Additions. Where *play areas* are designed and constructed in phases, the requirements of 240 shall apply to each successive *addition* so that when the *addition* is completed, the entire *play area* complies with all the applicable requirements of 240.

> **Advisory 240.1.1 Additions.** These requirements are to be applied so that when each successive addition is completed, the entire play area complies with all applicable provisions. For example, a play area is built in two phases. In the first phase, there are 10 elevated play components and 10 elevated play components are added in the second phase for a total of 20 elevated play components in the play area. When the first phase was completed, at least 5 elevated play components, including at least 3 different types, were to be provided on an accessible route. When the second phase is completed, at least 10 elevated play components must be located on an accessible route, and at least 7 ground level play components, including 4 different types, must be provided on an accessible route. At the time the second phase is complete, ramps must be used to connect at least 5 of the elevated play components and transfer systems are permitted to be used to connect the rest of the elevated play components required to be located on an accessible route.

240.2 Play Components. Where provided, *play components* shall comply with 240.2.

240.2.1 Ground Level Play Components. *Ground level play components* shall be provided in the number and types required by 240.2.1. *Ground level play components* that are provided to comply with 240.2.1.1 shall be permitted to satisfy the additional number required by 240.2.1.2 if the minimum required types of *play components* are satisfied. Where two or more required *ground level play components* are provided, they shall be dispersed throughout the *play area* and integrated with other *play components*.

> **Advisory 240.2.1 Ground Level Play Components.** Examples of ground level play components may include spring rockers, swings, diggers, and stand-alone slides. When distinguishing between the different types of ground level play components, consider the general experience provided by the play component. Examples of different types of experiences include, but are not limited to, rocking, swinging, climbing, spinning, and sliding.

Advisory 240.2.1 Ground Level Play Components (Continued). A spiral slide may provide a slightly different experience from a straight slide, but sliding is the general experience and therefore a spiral slide is not considered a different type of play component from a straight slide.

Ground level play components accessed by children with disabilities must be integrated into the play area. Designers should consider the optimal layout of ground level play components accessed by children with disabilities to foster interaction and socialization among all children. Grouping all ground level play components accessed by children with disabilities in one location is not considered integrated.

Where a stand-alone slide is provided, an accessible route must connect the base of the stairs at the entry point to the exit point of the slide. A ramp or transfer system to the top of the slide is not required. Where a sand box is provided, an accessible route must connect to the border of the sand box. Accessibility to the sand box would be enhanced by providing a transfer system into the sand or by providing a raised sand table with knee clearance complying with 1008.4.3.

Ramps are preferred over transfer systems since not all children who use wheelchairs or other mobility devices may be able to use, or may choose not to use, transfer systems. Where ramps connect elevated play components, the maximum rise of any ramp run is limited to 12 inches (305 mm). Where possible, designers and operators are encouraged to provide ramps with a slope less than the 1:12 maximum. Berms or sculpted dirt may be used to provide elevation and may be part of an accessible route to composite play structures.

Platform lifts are permitted as a part of an accessible route. Because lifts must be independently operable, operators should carefully consider the appropriateness of their use in unsupervised settings.

240.2.1.1 Minimum Number and Types. Where *ground level play components* are provided, at least one of each type shall be on an *accessible* route and shall comply with 1008.4.

240.2.1.2 Additional Number and Types. Where *elevated play components* are provided, *ground level play components* shall be provided in accordance with Table 240.2.1.2 and shall comply with 1008.4.

EXCEPTION: If at least 50 percent of the *elevated play components* are connected by a *ramp* and at least 3 of the *elevated play components* connected by the *ramp* are different types of *play components*, the *play area* shall not be required to comply with 240.2.1.2.

Table 240.2.1.2 Number and Types of Ground Level Play Components Required to be on Accessible Routes

Number of Elevated Play Components Provided	Minimum Number of Ground Level Play Components Required to be on an Accessible Route	Minimum Number of Different Types of Ground Level Play Components Required to be on an Accessible Route
1	Not applicable	Not applicable
2 to 4	1	1
5 to 7	2	2
8 to 10	3	3
11 to 13	4	3
14 to 16	5	3
17 to 19	6	3
20 to 22	7	4
23 to 25	8	4
26 and over	8, plus 1 for each additional 3, or fraction thereof, over 25	5

> **Advisory 240.2.1.2 Additional Number and Types.** Where a large play area includes two or more composite play structures designed for the same age group, the total number of elevated play components on all the composite play structures must be added to determine the additional number and types of ground level play components that must be provided on an accessible route.

240.2.2 Elevated Play Components. Where *elevated play components* are provided, at least 50 percent shall be on an *accessible* route and shall comply with 1008.4.

> **Advisory 240.2.2 Elevated Play Components.** A double or triple slide that is part of a composite play structure is one elevated play component. For purposes of this section, ramps, transfer systems, steps, decks, and roofs are not considered elevated play components. Although socialization and pretend play can occur on these elements, they are not primarily intended for play.
>
> Some play components that are attached to a composite play structure can be approached or exited at the ground level or above grade from a platform or deck. For example, a climber attached to a composite play structure can be approached or exited at the ground level or above grade from a platform or deck on a composite play structure.

> **Advisory 240.2.2 Elevated Play Components (Continued).** Play components that are attached to a composite play structure and can be approached from a platform or deck (e.g., climbers and overhead play components) are considered elevated play components. These play components are not considered ground level play components and do not count toward the requirements in 240.2.1.2 regarding the number of ground level play components that must be located on an accessible route.

241 Saunas and Steam Rooms

241 General. Where provided, saunas and steam rooms shall comply with 612.
EXCEPTION: Where saunas or steam rooms are clustered at a single location, no more than 5 percent of the saunas and steam rooms, but no fewer than one, of each type in each cluster shall be required to comply with 612.

242 Swimming Pools, Wading Pools, and Spas

242.1 General. Swimming pools, wading pools, and spas shall comply with 242.

242.2 Swimming Pools. At least two *accessible* means of entry shall be provided for swimming pools. *Accessible* means of entry shall be swimming pool lifts complying with 1009.2; sloped entries complying with 1009.3; transfer walls complying with 1009.4; transfer systems complying with 1009.5; and pool stairs complying with 1009.6. At least one *accessible* means of entry provided shall comply with 1009.2 or 1009.3.
EXCEPTIONS: 1. Where a swimming pool has less than 300 linear feet (91 m) of swimming pool wall, no more than one *accessible* means of entry shall be required provided that the *accessible* means of entry is a swimming pool lift complying with 1009.2 or sloped entry complying with 1009.3.
2. Wave action pools, leisure rivers, sand bottom pools, and other pools where user access is limited to one area shall not be required to provide more than one *accessible* means of entry provided that the *accessible* means of entry is a swimming pool lift complying with 1009.2, a sloped entry complying with 1009.3, or a transfer system complying with 1009.5.
3. *Catch pools* shall not be required to provide an *accessible* means of entry provided that the *catch pool* edge is on an *accessible* route.

> **Advisory 242.2 Swimming Pools.** Where more than one means of access is provided into the water, it is recommended that the means be different. Providing different means of access will better serve the varying needs of people with disabilities in getting into and out of a swimming pool. It is also recommended that where two or more means of access are provided, they not be provided in the same location in the pool. Different locations will provide increased options for entry and exit, especially in larger pools.
>
> **Advisory 242.2 Swimming Pools Exception 1.** Pool walls at diving areas and areas along pool walls where there is no pool entry because of landscaping or adjacent structures are to be counted when determining the number of accessible means of entry required.

242.3 Wading Pools. At least one *accessible* means of entry shall be provided for wading pools. *Accessible* means of entry shall comply with sloped entries complying with 1009.3.

242.4 Spas. At least one *accessible* means of entry shall be provided for spas. *Accessible* means of entry shall comply with swimming pool lifts complying with 1009.2; transfer walls complying with 1009.4; or transfer systems complying with 1009.5.

    EXCEPTION: Where spas are provided in a cluster, no more than 5 percent, but no fewer than one, spa in each cluster shall be required to comply with 242.4.

243 Shooting Facilities with Firing Positions

243.1 General. Where shooting *facilities* with firing positions are designed and constructed at a *site*, at least 5 percent, but no fewer than one, of each type of firing position shall comply with 1010.

CHAPTER 3: BUILDING BLOCKS

301 General

301.1 Scope. The provisions of Chapter 3 shall apply where required by Chapter 2 or where referenced by a requirement in this document.

302 Floor or Ground Surfaces

302.1 General. Floor and ground surfaces shall be stable, firm, and slip resistant and shall comply with 302.
    EXCEPTIONS: 1. Within animal containment areas, floor and ground surfaces shall not be required to be stable, firm, and slip resistant.
    2. *Areas of sport activity* shall not be required to comply with 302.

> **Advisory 302.1 General.** A stable surface is one that remains unchanged by contaminants or applied force, so that when the contaminant or force is removed, the surface returns to its original condition. A firm surface resists deformation by either indentations or particles moving on its surface. A slip-resistant surface provides sufficient frictional counterforce to the forces exerted in walking to permit safe ambulation.

302.2 Carpet. Carpet or carpet tile shall be securely attached and shall have a firm cushion, pad, or backing or no cushion or pad. Carpet or carpet tile shall have a level loop, textured loop, level cut pile, or level cut/uncut pile texture. Pile height shall be ½ inch (13 mm) maximum. Exposed edges of carpet shall be fastened to floor surfaces and shall have trim on the entire length of the exposed edge. Carpet edge trim shall comply with 303.

> **Advisory 302.2 Carpet.** Carpets and permanently affixed mats can significantly increase the amount of force (roll resistance) needed to propel a wheelchair over a surface. The firmer the carpeting and backing, the lower the roll resistance. A pile thickness up to ½ inch (13 mm) (measured to the backing, cushion, or pad) is allowed, although a lower pile provides easier wheelchair maneuvering. If a backing, cushion or pad is used, it must be firm. Preferably, carpet pad should not be used because the soft padding increases roll resistance.

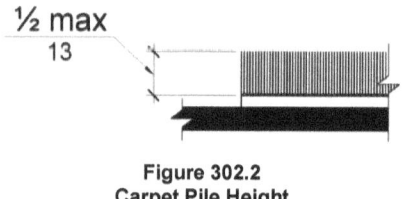

Figure 302.2
Carpet Pile Height

302.3 Openings. Openings in floor or ground surfaces shall not allow passage of a sphere more than ½ inch (13 mm) diameter except as allowed in 407.4.3, 409.4.3, 410.4, 810.5.3 and 810.10. Elongated openings shall be placed so that the long dimension is perpendicular to the dominant direction of travel.

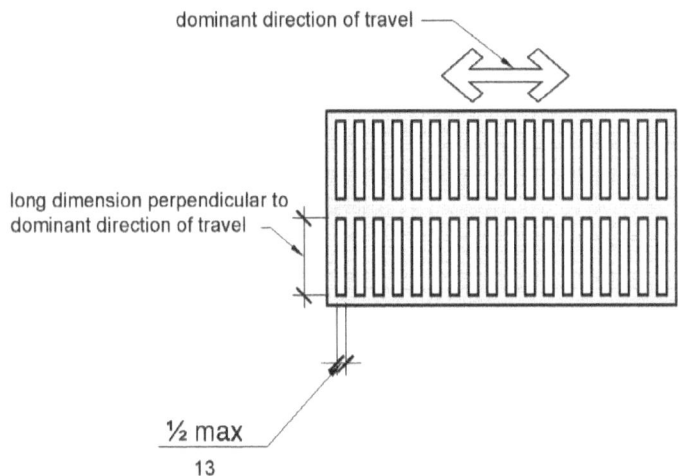

Figure 302.3
Elongated Openings in Floor or Ground Surfaces

303 Changes in Level

303.1 General. Where changes in level are permitted in floor or ground surfaces, they shall comply with 303.
   EXCEPTIONS: 1. Animal containment areas shall not be required to comply with 303.
   2. *Areas of sport activity* shall not be required to comply with 303.

303.2 Vertical. Changes in level of ¼ inch (6.4 mm) high maximum shall be permitted to be vertical.

Figure 303.2
Vertical Change in Level

303.3 Beveled. Changes in level between ¼ inch (6.4 mm) high minimum and ½ inch (13 mm) high maximum shall be beveled with a slope not steeper than 1:2.

> **Advisory 303.3 Beveled.** A change in level of ½ inch (13 mm) is permitted to be ¼ inch (6.4 mm) vertical plus ¼ inch (6.4 mm) beveled. However, in no case may the combined change in level exceed ½ inch (13 mm). Changes in level exceeding ½ inch (13 mm) must comply with 405 (Ramps) or 406 (Curb Ramps).

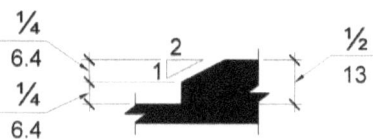

Figure 303.3
Beveled Change in Level

303.4 Ramps. Changes in level greater than ½ inch (13 mm) high shall be *ramped*, and shall comply with 405 or 406.

304 Turning Space

304.1 General. Turning *space* shall comply with 304.

304.2 Floor or Ground Surfaces. Floor or ground surfaces of a turning *space* shall comply with 302. Changes in level are not permitted.
  EXCEPTION: Slopes not steeper than 1:48 shall be permitted.

> **Advisory 304.2 Floor or Ground Surface Exception.** As used in this section, the phrase "changes in level" refers to surfaces with slopes and to surfaces with abrupt rise exceeding that permitted in Section 303.3. Such changes in level are prohibited in required clear floor and ground spaces, turning spaces, and in similar spaces where people using wheelchairs and other mobility devices must park their mobility aids such as in wheelchair spaces, or maneuver to use elements such as at doors, fixtures, and telephones. The exception permits slopes not steeper than 1:48.

304.3 Size. Turning *space* shall comply with 304.3.1 or 304.3.2.

  304.3.1 Circular Space. The turning *space* shall be a *space* of 60 inches (1525 mm) diameter minimum. The *space* shall be permitted to include knee and toe clearance complying with 306.

  304.3.2 T-Shaped Space. The turning *space* shall be a T-shaped *space* within a 60 inch (1525 mm) square minimum with arms and base 36 inches (915 mm) wide minimum. Each arm of the T shall be clear of obstructions 12 inches (305 mm) minimum in each direction and the base shall be clear of obstructions 24 inches (610 mm) minimum. The *space* shall be permitted to include knee and toe clearance complying with 306 only at the end of either the base or one arm.

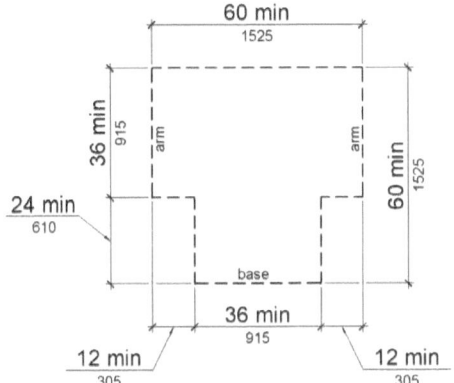

Figure 304.3.2
T-Shaped Turning Space

304.4 Door Swing. Doors shall be permitted to swing into turning *spaces*.

305 Clear Floor or Ground Space

305.1 General. Clear floor or ground *space* shall comply with 305.

305.2 Floor or Ground Surfaces. Floor or ground surfaces of a clear floor or ground *space* shall comply with 302. Changes in level are not permitted.
  EXCEPTION: Slopes not steeper than 1:48 shall be permitted.

305.3 Size. The clear floor or ground *space* shall be 30 inches (760 mm) minimum by 48 inches (1220 mm) minimum.

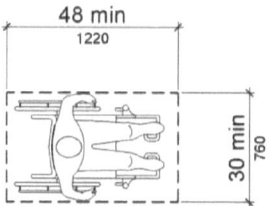

Figure 305.3
Clear Floor or Ground Space

305.4 Knee and Toe Clearance. Unless otherwise specified, clear floor or ground *space* shall be permitted to include knee and toe clearance complying with 306.

305.5 Position. Unless otherwise specified, clear floor or ground *space* shall be positioned for either forward or parallel approach to an *element*.

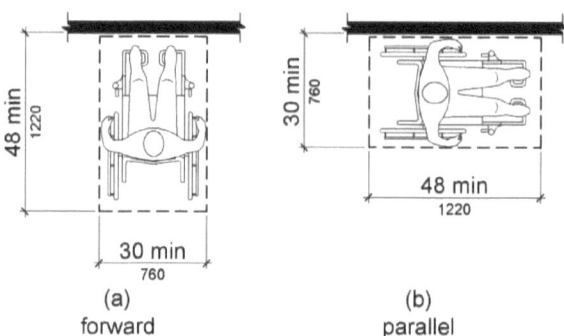

(a) forward

(b) parallel

Figure 305.5
Position of Clear Floor or Ground Space

305.6 Approach. One full unobstructed side of the clear floor or ground *space* shall adjoin an *accessible* route or adjoin another clear floor or ground *space*.

305.7 Maneuvering Clearance. Where a clear floor or ground *space* is located in an alcove or otherwise confined on all or part of three sides, additional maneuvering clearance shall be provided in accordance with 305.7.1 and 305.7.2.

305.7.1 Forward Approach. Alcoves shall be 36 inches (915 mm) wide minimum where the depth exceeds 24 inches (610 mm).

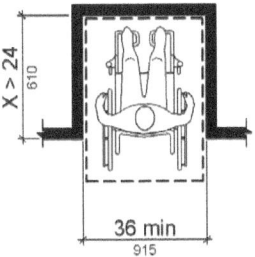

Figure 305.7.1
Maneuvering Clearance in an Alcove, Forward Approach

305.7.2 Parallel Approach. Alcoves shall be 60 inches (1525 mm) wide minimum where the depth exceeds 15 inches (380 mm).

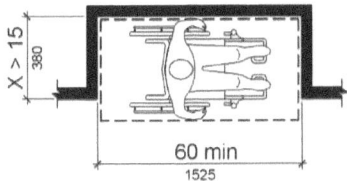

Figure 305.7.2
Maneuvering Clearance in an Alcove, Parallel Approach

306 Knee and Toe Clearance

306.1 General. Where *space* beneath an *element* is included as part of clear floor or ground *space* or turning *space*, the *space* shall comply with 306. Additional *space* shall not be prohibited beneath an *element* but shall not be considered as part of the clear floor or ground *space* or turning *space*.

> **Advisory 306.1 General.** Clearances are measured in relation to the usable clear floor space, not necessarily to the vertical support for an element. When determining clearance under an object for required turning or maneuvering space, care should be taken to ensure the space is clear of any obstructions.

306.2 Toe Clearance.

306.2.1 General. *Space* under an *element* between the finish floor or ground and 9 inches (230 mm) above the finish floor or ground shall be considered toe clearance and shall comply with 306.2.

306.2.2 Maximum Depth. Toe clearance shall extend 25 inches (635 mm) maximum under an *element*.

306.2.3 Minimum Required Depth. Where toe clearance is required at an *element* as part of a clear floor *space*, the toe clearance shall extend 17 inches (430 mm) minimum under the *element*.

306.2.4 Additional Clearance. *Space* extending greater than 6 inches (150 mm) beyond the available knee clearance at 9 inches (230 mm) above the finish floor or ground shall not be considered toe clearance.

306.2.5 Width. Toe clearance shall be 30 inches (760 mm) wide minimum.

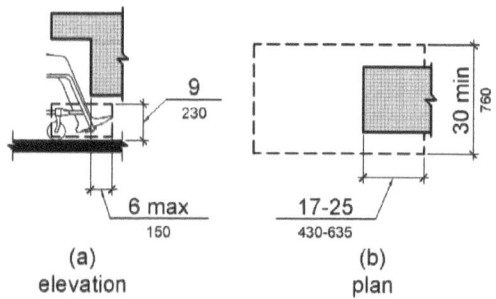

(a) elevation

(b) plan

Figure 306.2
Toe Clearance

306.3 Knee Clearance.

306.3.1 General. *Space* under an *element* between 9 inches (230 mm) and 27 inches (685 mm) above the finish floor or ground shall be considered knee clearance and shall comply with 306.3.

306.3.2 Maximum Depth. Knee clearance shall extend 25 inches (635 mm) maximum under an *element* at 9 inches (230 mm) above the finish floor or ground.

306.3.3 Minimum Required Depth. Where knee clearance is required under an *element* as part of a clear floor *space*, the knee clearance shall be 11 inches (280 mm) deep minimum at 9 inches (230 mm) above the finish floor or ground, and 8 inches (205 mm) deep minimum at 27 inches (685 mm) above the finish floor or ground.

306.3.4 Clearance Reduction. Between 9 inches (230 mm) and 27 inches (685 mm) above the finish floor or ground, the knee clearance shall be permitted to reduce at a rate of 1 inch (25 mm) in depth for each 6 inches (150 mm) in height.

306.3.5 Width. Knee clearance shall be 30 inches (760 mm) wide minimum.

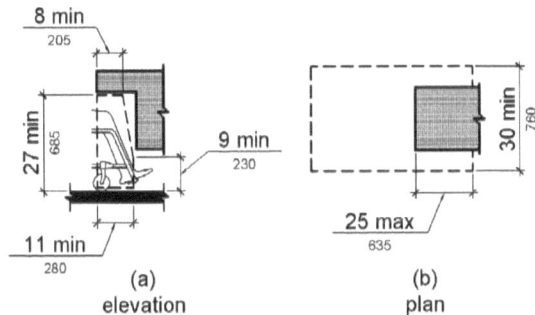

Figure 306.3
Knee Clearance

307 Protruding Objects

307.1 General. Protruding objects shall comply with 307.

307.2 Protrusion Limits. Objects with leading edges more than 27 inches (685 mm) and not more than 80 inches (2030 mm) above the finish floor or ground shall protrude 4 inches (100 mm) maximum horizontally into the *circulation path*.
   EXCEPTION: Handrails shall be permitted to protrude 4½ inches (115 mm) maximum.

> **Advisory 307.2 Protrusion Limits.** When a cane is used and the element is in the detectable range, it gives a person sufficient time to detect the element with the cane before there is body contact. Elements located on circulation paths, including operable elements, must comply with requirements for protruding objects. For example, awnings and their supporting structures cannot reduce the minimum required vertical clearance. Similarly, casement windows, when open, cannot encroach more than 4 inches (100 mm) into circulation paths above 27 inches (685 mm).

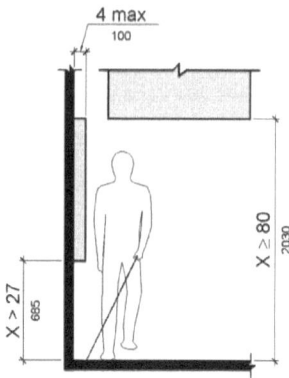

Figure 307.2
Limits of Protruding Objects

307.3 Post-Mounted Objects. Free-standing objects mounted on posts or pylons shall overhang *circulation paths* 12 inches (305 mm) maximum when located 27 inches (685 mm) minimum and 80 inches (2030 mm) maximum above the finish floor or ground. Where a sign or other obstruction is mounted between posts or pylons and the clear distance between the posts or pylons is greater than 12 inches (305 mm), the lowest edge of such sign or obstruction shall be 27 inches (685 mm) maximum or 80 inches (2030 mm) minimum above the finish floor or ground.

  EXCEPTION: The sloping portions of handrails serving stairs and *ramps* shall not be required to comply with 307.3.

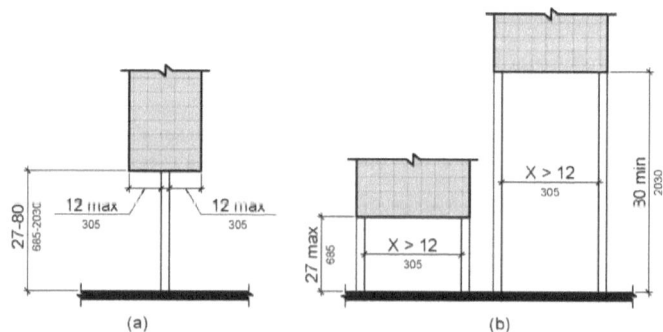

Figure 307.3
Post-Mounted Protruding Objects

307.4 Vertical Clearance. Vertical clearance shall be 80 inches (2030 mm) high minimum. Guardrails or other barriers shall be provided where the vertical clearance is less than 80 inches (2030 mm) high. The leading edge of such guardrail or barrier shall be located 27 inches (685 mm) maximum above the finish floor or ground.
    EXCEPTION: Door closers and door stops shall be permitted to be 78 inches (1980 mm) minimum above the finish floor or ground.

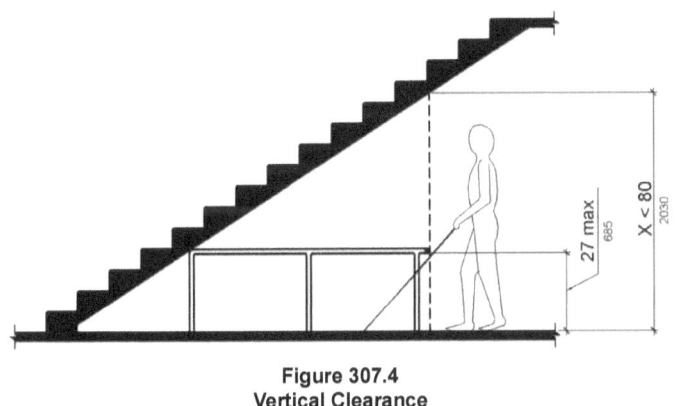

Figure 307.4
Vertical Clearance

307.5 Required Clear Width. Protruding objects shall not reduce the clear width required for *accessible* routes.

308 Reach Ranges

308.1 General. Reach ranges shall comply with 308.

> **Advisory 308.1 General.** The following table provides guidance on reach ranges for children according to age where building elements such as coat hooks, lockers, or operable parts are designed for use primarily by children. These dimensions apply to either forward or side reaches. Accessible elements and operable parts designed for adult use or children over age 12 can be located outside these ranges but must be within the adult reach ranges required by 308.
>
> | Children's Reach Ranges | | | |
> |---|---|---|---|
> | Forward or Side Reach | Ages 3 and 4 | Ages 5 through 8 | Ages 9 through 12 |
> | High (maximum) | 36 in (915 mm) | 40 in (1015 mm) | 44 in (1120 mm) |
> | Low (minimum) | 20 in (510 mm) | 18 in (455 mm) | 16 in (405 mm) |

308.2 Forward Reach.

308.2.1 Unobstructed. Where a forward reach is unobstructed, the high forward reach shall be 48 inches (1220 mm) maximum and the low forward reach shall be 15 inches (380 mm) minimum above the finish floor or ground.

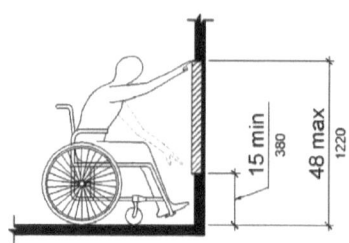

Figure 308.2.1
Unobstructed Forward Reach

308.2.2 Obstructed High Reach. Where a high forward reach is over an obstruction, the clear floor *space* shall extend beneath the *element* for a distance not less than the required reach depth over the obstruction. The high forward reach shall be 48 inches (1220 mm) maximum where the reach depth is 20 inches (510 mm) maximum. Where the reach depth exceeds 20 inches (510 mm), the high forward reach shall be 44 inches (1120 mm) maximum and the reach depth shall be 25 inches (635 mm) maximum.

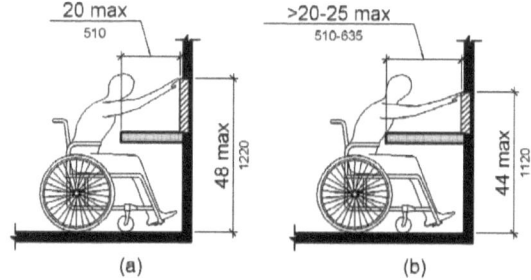

Figure 308.2.2
Obstructed High Forward Reach

308.3 Side Reach.

308.3.1 Unobstructed. Where a clear floor or ground *space* allows a parallel approach to an *element* and the side reach is unobstructed, the high side reach shall be 48 inches (1220 mm)

maximum and the low side reach shall be 15 inches (380 mm) minimum above the finish floor or ground.
    EXCEPTIONS: 1. An obstruction shall be permitted between the clear floor or ground *space* and the *element* where the depth of the obstruction is 10 inches (255 mm) maximum.
    2. *Operable parts* of fuel dispensers shall be permitted to be 54 inches (1370 mm) maximum measured from the surface of the *vehicular way* where fuel dispensers are installed on existing curbs.

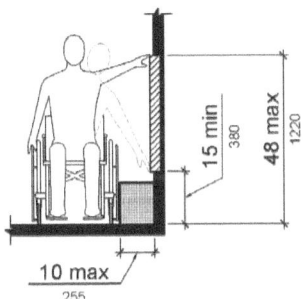

**Figure 308.3.1
Unobstructed Side Reach**

308.3.2 Obstructed High Reach. Where a clear floor or ground *space* allows a parallel approach to an *element* and the high side reach is over an obstruction, the height of the obstruction shall be 34 inches (865 mm) maximum and the depth of the obstruction shall be 24 inches (610 mm) maximum. The high side reach shall be 48 inches (1220 mm) maximum for a reach depth of 10 inches (255 mm) maximum. Where the reach depth exceeds 10 inches (255 mm), the high side reach shall be 46 inches (1170 mm) maximum for a reach depth of 24 inches (610 mm) maximum.
    EXCEPTIONS: 1. The top of washing machines and clothes dryers shall be permitted to be 36 inches (915 mm) maximum above the finish floor.
    2. *Operable parts* of fuel dispensers shall be permitted to be 54 inches (1370 mm) maximum measured from the surface of the *vehicular way* where fuel dispensers are installed on existing curbs.

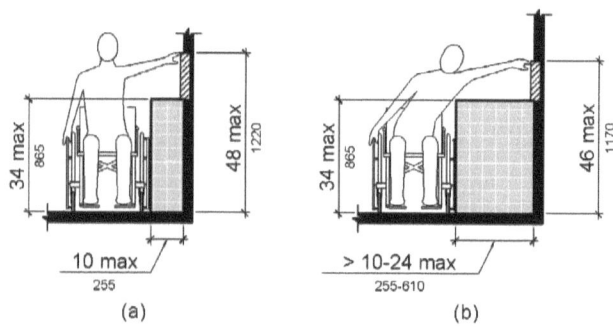

Figure 308.3.2
Obstructed High Side Reach

309 Operable Parts

309.1 General. *Operable parts* shall comply with 309.

309.2 Clear Floor Space. A clear floor or ground *space* complying with 305 shall be provided.

309.3 Height. *Operable parts* shall be placed within one or more of the reach ranges specified in 308.

309.4 Operation. *Operable parts* shall be operable with one hand and shall not require tight grasping, pinching, or twisting of the wrist. The force required to activate *operable parts* shall be 5 pounds (22.2 N) maximum.
   EXCEPTION: Gas pump nozzles shall not be required to provide *operable parts* that have an activating force of 5 pounds (22.2 N) maximum.

CHAPTER 4: ACCESSIBLE ROUTES

401 General

401.1 Scope. The provisions of Chapter 4 shall apply where required by Chapter 2 or where referenced by a requirement in this document.

402 Accessible Routes

402.1 General. *Accessible* routes shall comply with 402.

402.2 Components. *Accessible* routes shall consist of one or more of the following components: walking surfaces with a *running slope* not steeper than 1:20, doorways, *ramps*, *curb ramps* excluding the flared sides, elevators, and platform lifts. All components of an *accessible* route shall comply with the applicable requirements of Chapter 4.

> **Advisory 402.2 Components.** Walking surfaces must have running slopes not steeper than 1:20, see 403.3. Other components of accessible routes, such as ramps (405) and curb ramps (406), are permitted to be more steeply sloped.

403 Walking Surfaces

403.1 General. Walking surfaces that are a part of an *accessible* route shall comply with 403.

403.2 Floor or Ground Surface. Floor or ground surfaces shall comply with 302.

403.3 Slope. The *running slope* of walking surfaces shall not be steeper than 1:20. The *cross slope* of walking surfaces shall not be steeper than 1:48.

403.4 Changes in Level. Changes in level shall comply with 303.

403.5 Clearances. Walking surfaces shall provide clearances complying with 403.5.
EXCEPTION: Within *employee work areas*, clearances on *common use circulation paths* shall be permitted to be decreased by *work area equipment* provided that the decrease is essential to the function of the work being performed.

    403.5.1 Clear Width. Except as provided in 403.5.2 and 403.5.3, the clear width of walking surfaces shall be 36 inches (915 mm) minimum.
    EXCEPTION: The clear width shall be permitted to be reduced to 32 inches (815 mm) minimum for a length of 24 inches (610 mm) maximum provided that reduced width segments are separated by segments that are 48 inches (1220 mm) long minimum and 36 inches (915 mm) wide minimum.

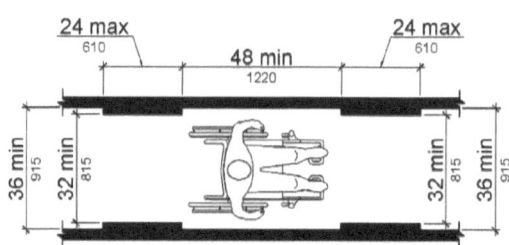

Figure 403.5.1
Clear Width of an Accessible Route

403.5.2 Clear Width at Turn. Where the *accessible* route makes a 180 degree turn around an *element* which is less than 48 inches (1220 mm) wide, clear width shall be 42 inches (1065 mm) minimum approaching the turn, 48 inches (1220 mm) minimum at the turn and 42 inches (1065 mm) minimum leaving the turn.

EXCEPTION: Where the clear width at the turn is 60 inches (1525 mm) minimum compliance with 403.5.2 shall not be required.

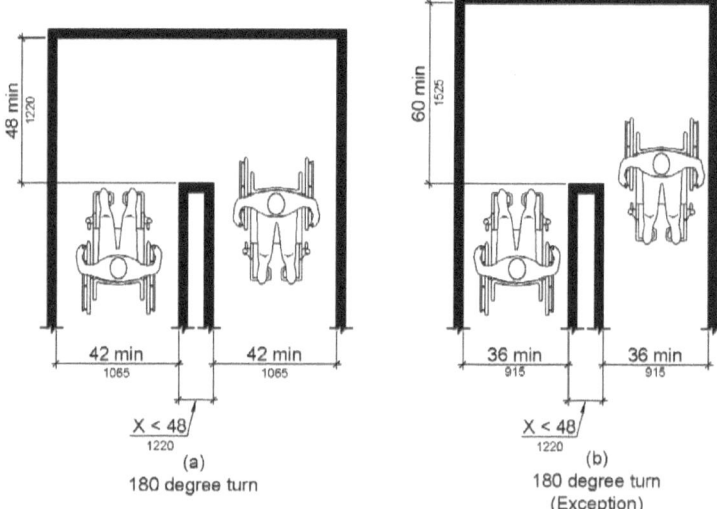

Figure 403.5.2
Clear Width at Turn

403.5.3 Passing Spaces. An *accessible* route with a clear width less than 60 inches (1525 mm) shall provide passing *spaces* at intervals of 200 feet (61 m) maximum. Passing *spaces* shall be either: a *space* 60 inches (1525 mm) minimum by 60 inches (1525 mm) minimum; or, an intersection of two walking surfaces providing a T-shaped *space* complying with 304.3.2 where the base and arms of the T-shaped *space* extend 48 inches (1220 mm) minimum beyond the intersection.

403.6 Handrails. Where handrails are provided along walking surfaces with *running slopes* not steeper than 1:20 they shall comply with 505.

> **Advisory 403.6 Handrails.** Handrails provided in elevator cabs and platform lifts are not required to comply with the requirements for handrails on walking surfaces.

404 Doors, Doorways, and Gates

404.1 General. Doors, doorways, and gates that are part of an *accessible* route shall comply with 404.
EXCEPTION: Doors, doorways, and gates designed to be operated only by security personnel shall not be required to comply with 404.2.7, 404.2.8, 404.2.9, 404.3.2 and 404.3.4 through 404.3.7.

> **Advisory 404.1 General Exception.** Security personnel must have sole control of doors that are eligible for the Exception at 404.1. It would not be acceptable for security personnel to operate the doors for people with disabilities while allowing others to have independent access.

404.2 Manual Doors, Doorways, and Manual Gates. Manual doors and doorways and manual gates intended for user passage shall comply with 404.2.

404.2.1 Revolving Doors, Gates, and Turnstiles. Revolving doors, revolving gates, and turnstiles shall not be part of an *accessible* route.

404.2.2 Double-Leaf Doors and Gates. At least one of the active leaves of doorways with two leaves shall comply with 404.2.3 and 404.2.4.

404.2.3 Clear Width. Door openings shall provide a clear width of 32 inches (815 mm) minimum. Clear openings of doorways with swinging doors shall be measured between the face of the door and the stop, with the door open 90 degrees. Openings more than 24 inches (610 mm) deep shall provide a clear opening of 36 inches (915 mm) minimum. There shall be no projections into the required clear opening width lower than 34 inches (865 mm) above the finish floor or ground. Projections into the clear opening width between 34 inches (865 mm) and 80 inches (2030 mm) above the finish floor or ground shall not exceed 4 inches (100 mm).
    EXCEPTIONS: 1. In *alterations*, a projection of 5/8 inch (16 mm) maximum into the required clear width shall be permitted for the latch side stop.
    2. Door closers and door stops shall be permitted to be 78 inches (1980 mm) minimum above the finish floor or ground.

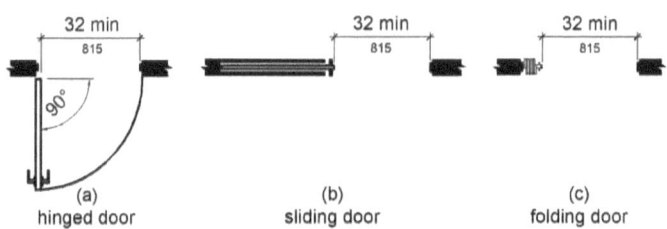

**Figure 404.2.3
Clear Width of Doorways**

404.2.4 Maneuvering Clearances. Minimum maneuvering clearances at doors and gates shall comply with 404.2.4. Maneuvering clearances shall extend the full width of the doorway and the required latch side or hinge side clearance.

EXCEPTION: Entry doors to hospital patient rooms shall not be required to provide the clearance beyond the latch side of the door.

404.2.4.1 Swinging Doors and Gates. Swinging doors and gates shall have maneuvering clearances complying with Table 404.2.4.1.

Table 404.2.4.1 Maneuvering Clearances at Manual Swinging Doors and Gates

Type of Use		Minimum Maneuvering Clearance	
Approach Direction	Door or Gate Side	Perpendicular to Doorway	Parallel to Doorway (beyond latch side unless noted)
From front	Pull	60 inches (1525 mm)	18 inches (455 mm)
From front	Push	48 inches (1220 mm)	0 inches (0 mm)[1]
From hinge side	Pull	60 inches (1525 mm)	36 inches (915 mm)
From hinge side	Pull	54 inches (1370 mm)	42 inches (1065 mm)
From hinge side	Push	42 inches (1065 mm)[2]	22 inches (560 mm)[3]
From latch side	Pull	48 inches (1220 mm)[4]	24 inches (610 mm)
From latch side	Push	42 inches (1065 mm)[4]	24 inches (610 mm)

1. Add 12 inches (305 mm) if closer and latch are provided.
2. Add 6 inches (150 mm) if closer and latch are provided.
3. Beyond hinge side.
4. Add 6 inches (150 mm) if closer is provided.

CHAPTER 4: ACCESSIBLE ROUTES

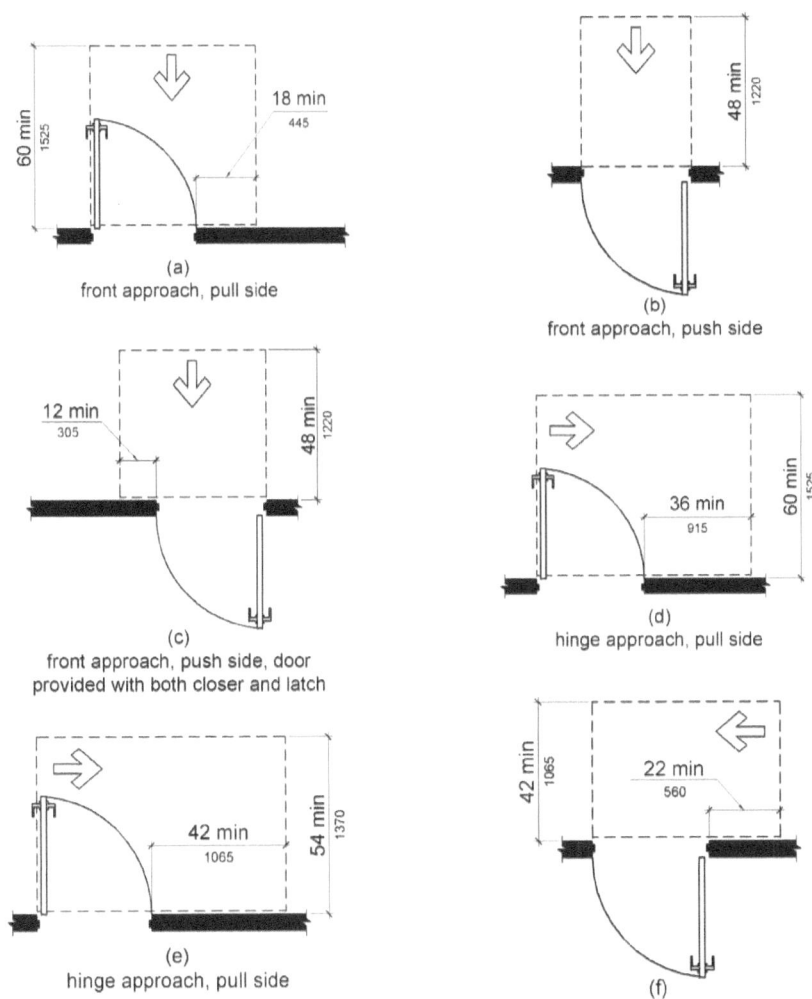

Figure 404.2.4.1
Maneuvering Clearances at Manual Swinging Doors and Gates

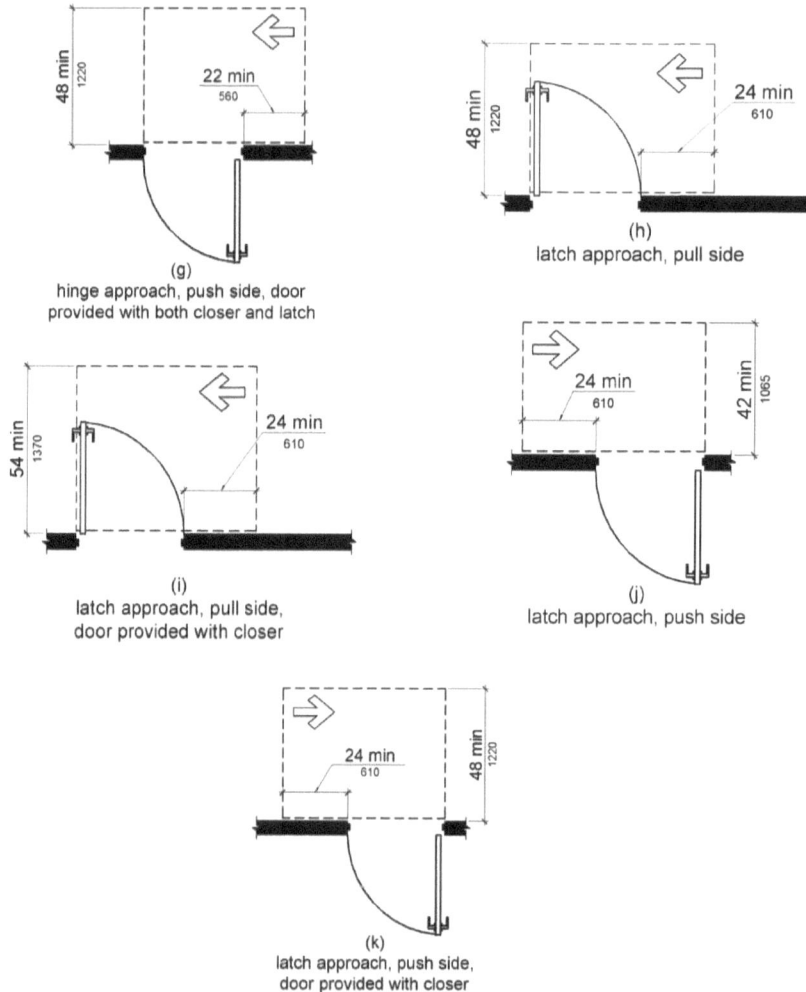

Figure 404.2.4.1
Maneuvering Clearances at Manual Swinging Doors and Gates

404.2.4.2 Doorways without Doors or Gates, Sliding Doors, and Folding Doors. Doorways less than 36 inches (915 mm) wide without doors or gates, sliding doors, or folding doors shall have maneuvering clearances complying with Table 404.2.4.2.

Table 404.2.4.2 Maneuvering Clearances at Doorways without Doors or Gates, Manual Sliding Doors, and Manual Folding Doors

Approach Direction	Minimum Maneuvering Clearance	
	Perpendicular to Doorway	Parallel to Doorway (beyond stop/latch side unless noted)
From Front	48 inches (1220 mm)	0 inches (0 mm)
From side[1]	42 inches (1065 mm)	0 inches (0 mm)
From pocket/hinge side	42 inches (1065 mm)	22 inches (560 mm)[2]
From stop/latch side	42 inches (1065 mm)	24 inches (610 mm)

1. Doorway with no door only.
2. Beyond pocket/hinge side.

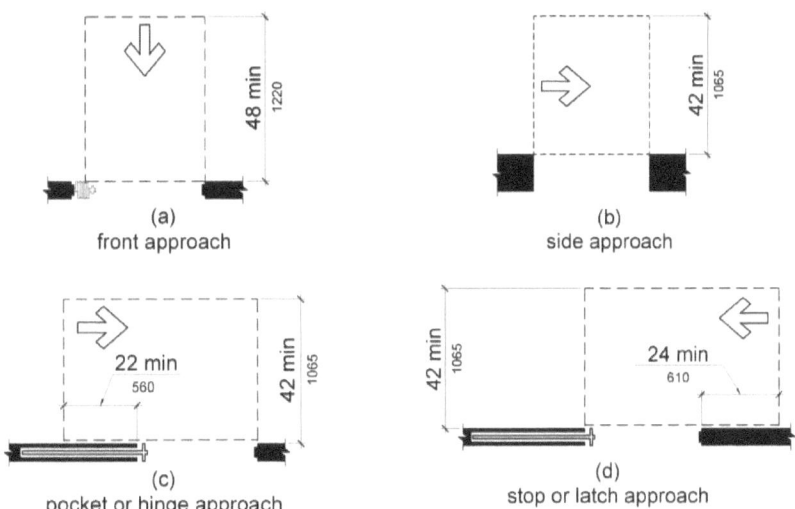

Figure 404.2.4.2
Maneuvering Clearances at Doorways without Doors, Sliding Doors, Gates, and Folding Doors

404.2.4.3 Recessed Doors and Gates. Maneuvering clearances for forward approach shall be provided when any obstruction within 18 inches (455 mm) of the latch side of a doorway projects more than 8 inches (205 mm) beyond the face of the door, measured perpendicular to the face of the door or gate.

> **Advisory 404.2.4.3 Recessed Doors and Gates.** A door can be recessed due to wall thickness or because of the placement of casework and other fixed elements adjacent to the doorway. This provision must be applied wherever doors are recessed.

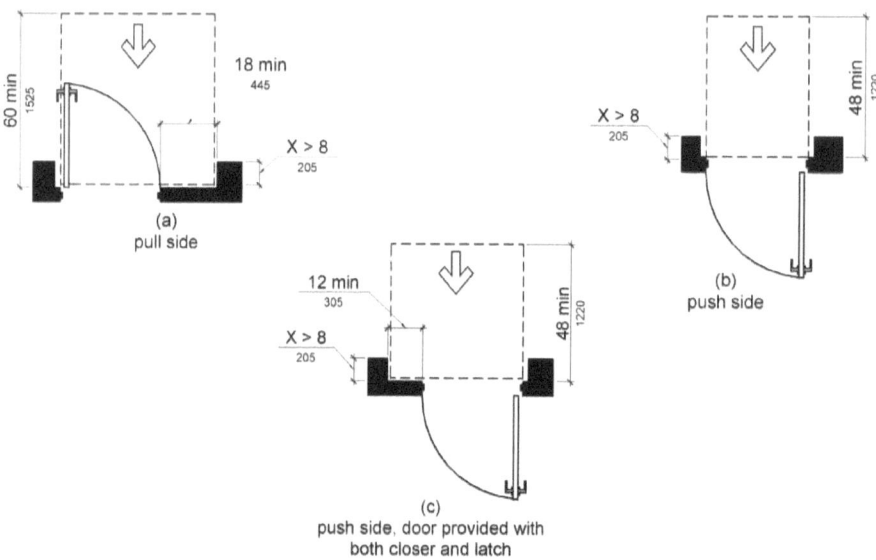

Figure 404.2.4.3
Maneuvering Clearances at Recessed Doors and Gates

404.2.4.4 Floor or Ground Surface. Floor or ground surface within required maneuvering clearances shall comply with 302. Changes in level are not permitted.
  EXCEPTIONS: **1.** Slopes not steeper than 1:48 shall be permitted.
  2. Changes in level at thresholds complying with 404.2.5 shall be permitted.

404.2.5 Thresholds. Thresholds, if provided at doorways, shall be ½ inch (13 mm) high maximum. Raised thresholds and changes in level at doorways shall comply with 302 and 303.
  EXCEPTION: Existing or *altered* thresholds ¾ inch (19 mm) high maximum that have a beveled edge on each side with a slope not steeper than 1:2 shall not be required to comply with 404.2.5.

404.2.6 Doors in Series and Gates in Series. The distance between two hinged or pivoted doors in series and gates in series shall be 48 inches (1220 mm) minimum plus the width of doors or gates swinging into the *space*.

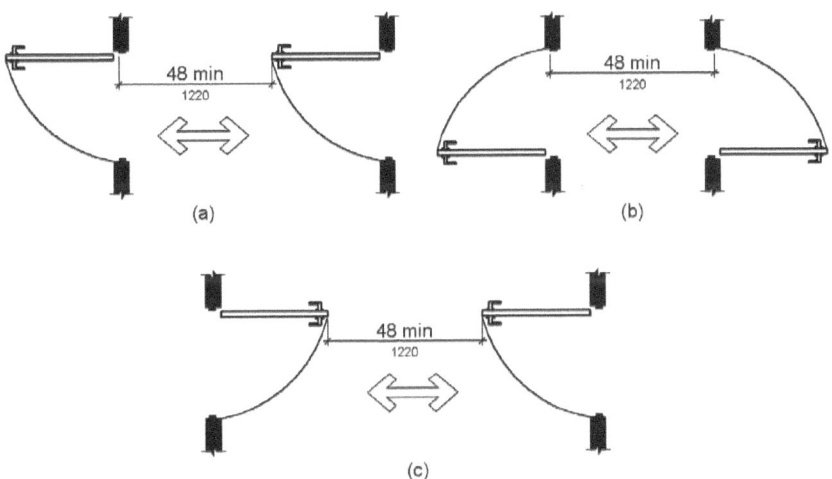

Figure 404.2.6
Doors in Series and Gates in Series

404.2.7 Door and Gate Hardware. Handles, pulls, latches, locks, and other *operable parts* on doors and gates shall comply with 309.4. *Operable parts* of such hardware shall be 34 inches (865 mm) minimum and 48 inches (1220 mm) maximum above the finish floor or ground. Where sliding doors are in the fully open position, operating hardware shall be exposed and usable from both sides.
EXCEPTIONS: 1. Existing locks shall be permitted in any location at existing glazed doors without stiles, existing overhead rolling doors or grilles, and similar existing doors or grilles that are designed with locks that are activated only at the top or bottom rail.
2. Access gates in barrier walls and fences protecting pools, spas, and hot tubs shall be permitted to have *operable parts* of the release of latch on self-latching devices at 54 inches (1370 mm) maximum above the finish floor or ground provided the self-latching devices are not also self-locking devices and operated by means of a key, electronic opener, or integral combination lock.

> **Advisory 404.2.7 Door and Gate Hardware.** Door hardware that can be operated with a closed fist or a loose grip accommodates the greatest range of users. Hardware that requires simultaneous hand and finger movements require greater dexterity and coordination, and is not recommended.

404.2.8 Closing Speed. Door and gate closing speed shall comply with 404.2.8.

404.2.8.1 Door Closers and Gate Closers. Door closers and gate closers shall be adjusted so that from an open position of 90 degrees, the time required to move the door to a position of 12 degrees from the latch is 5 seconds minimum.

404.2.8.2 Spring Hinges. Door and gate spring hinges shall be adjusted so that from the open position of 70 degrees, the door or gate shall move to the closed position in 1.5 seconds minimum.

404.2.9 Door and Gate Opening Force. Fire doors shall have a minimum opening force allowable by the appropriate *administrative authority*. The force for pushing or pulling open a door or gate other than fire doors shall be as follows:
1. Interior hinged doors and gates: 5 pounds (22.2 N) maximum.
2. Sliding or folding doors: 5 pounds (22.2 N) maximum.

These forces do not apply to the force required to retract latch bolts or disengage other devices that hold the door or gate in a closed position.

> **Advisory 404.2.9 Door and Gate Opening Force.** The maximum force pertains to the continuous application of force necessary to fully open a door, not the initial force needed to overcome the inertia of the door. It does not apply to the force required to retract bolts or to disengage other devices used to keep the door in a closed position.

404.2.10 Door and Gate Surfaces. Swinging door and gate surfaces within 10 inches (255 mm) of the finish floor or ground measured vertically shall have a smooth surface on the push side extending the full width of the door or gate. Parts creating horizontal or vertical joints in these surfaces shall be within 1/16 inch (1.6 mm) of the same plane as the other. Cavities created by added kick plates shall be capped.

EXCEPTIONS: 1. Sliding doors shall not be required to comply with 404.2.10.
2. Tempered glass doors without stiles and having a bottom rail or shoe with the top leading edge tapered at 60 degrees minimum from the horizontal shall not be required to meet the 10 inch (255 mm) bottom smooth surface height requirement.
3. Doors and gates that do not extend to within 10 inches (255 mm) of the finish floor or ground shall not be required to comply with 404.2.10.
4. Existing doors and gates without smooth surfaces within 10 inches (255 mm) of the finish floor or ground shall not be required to provide smooth surfaces complying with 404.2.10 provided that if added kick plates are installed, cavities created by such kick plates are capped.

404.2.11 Vision Lights. Doors, gates, and side lights adjacent to doors or gates, containing one or more glazing panels that permit viewing through the panels shall have the bottom of at least one glazed panel located 43 inches (1090 mm) maximum above the finish floor.

EXCEPTION: Vision lights with the lowest part more than 66 inches (1675 mm) from the finish floor or ground shall not be required to comply with 404.2.11.

404.3 Automatic and Power-Assisted Doors and Gates. Automatic doors and automatic gates shall comply with 404.3. Full-powered automatic doors shall comply with ANSI/BHMA A156.10 (incorporated by reference, see "Referenced Standards" in Chapter 1). Low-energy and power-assisted doors shall comply with ANSI/BHMA A156.19 (1997 or 2002 edition) (incorporated by reference, see "Referenced Standards" in Chapter 1).

404.3.1 Clear Width. Doorways shall provide a clear opening of 32 inches (815 mm) minimum in power-on and power-off mode. The minimum clear width for automatic door systems in a doorway shall be based on the clear opening provided by all leaves in the open position.

404.3.2 Maneuvering Clearance. Clearances at power-assisted doors and gates shall comply with 404.2.4. Clearances at automatic doors and gates without standby power and serving an *accessible means of egress* shall comply with 404.2.4.
EXCEPTION: Where automatic doors and gates remain open in the power-off condition, compliance with 404.2.4 shall not be required.

404.3.3 Thresholds. Thresholds and changes in level at doorways shall comply with 404.2.5.

404.3.4 Doors in Series and Gates in Series. Doors in series and gates in series shall comply with 404.2.6.

404.3.5 Controls. Manually operated controls shall comply with 309. The clear floor *space* adjacent to the control shall be located beyond the arc of the door swing.

404.3.6 Break Out Opening. Where doors and gates without standby power are a part of a means of egress, the clear break out opening at swinging or sliding doors and gates shall be 32 inches (815 mm) minimum when operated in emergency mode.
EXCEPTION: Where manual swinging doors and gates comply with 404.2 and serve the same means of egress compliance with 404.3.6 shall not be required.

404.3.7 Revolving Doors, Revolving Gates, and Turnstiles. Revolving doors, revolving gates, and turnstiles shall not be part of an *accessible* route.

405 Ramps

405.1 General. *Ramps* on *accessible* routes shall comply with 405.
EXCEPTION: In *assembly areas*, aisle *ramps* adjacent to seating and not serving *elements* required to be on an *accessible* route shall not be required to comply with 405.

405.2 Slope. *Ramp* runs shall have a *running slope* not steeper than 1:12.
EXCEPTION: In existing *sites*, *buildings*, and *facilities*, *ramps* shall be permitted to have *running slopes* steeper than 1:12 complying with Table 405.2 where such slopes are necessary due to *space* limitations.

Table 405.2 Maximum Ramp Slope and Rise for Existing Sites, Buildings, and Facilities

Slope[1]	Maximum Rise
Steeper than 1:10 but not steeper than 1:8	3 inches (75 mm)
Steeper than 1:12 but not steeper than 1:10	6 inches (150 mm)

1. A slope steeper than 1:8 is prohibited.

> **Advisory 405.2 Slope.** To accommodate the widest range of users, provide ramps with the least possible running slope and, wherever possible, accompany ramps with stairs for use by those individuals for whom distance presents a greater barrier than steps, e.g., people with heart disease or limited stamina.

405.3 Cross Slope. *Cross slope* of *ramp* runs shall not be steeper than 1:48.

> **Advisory 405.3 Cross Slope.** Cross slope is the slope of the surface perpendicular to the direction of travel. Cross slope is measured the same way as slope is measured (i.e., the rise over the run).

405.4 Floor or Ground Surfaces. Floor or ground surfaces of *ramp* runs shall comply with 302. Changes in level other than the *running slope* and *cross slope* are not permitted on *ramp* runs.

405.5 Clear Width. The clear width of a *ramp* run and, where handrails are provided, the clear width between handrails shall be 36 inches (915 mm) minimum.

    EXCEPTION: Within *employee work area*s, the required clear width of *ramps* that are a part of *common use circulation path*s shall be permitted to be decreased by *work area equipment* provided that the decrease is essential to the function of the work being performed.

405.6 Rise. The rise for any *ramp* run shall be 30 inches (760 mm) maximum.

405.7 Landings. *Ramps* shall have landings at the top and the bottom of each *ramp* run. Landings shall comply with 405.7.

> **Advisory 405.7 Landings.** Ramps that do not have level landings at changes in direction can create a compound slope that will not meet the requirements of this document. Circular or curved ramps continually change direction. Curvilinear ramps with small radii also can create compound cross slopes and cannot, by their nature, meet the requirements for accessible routes. A level landing is needed at the accessible door to permit maneuvering and simultaneously door operation.

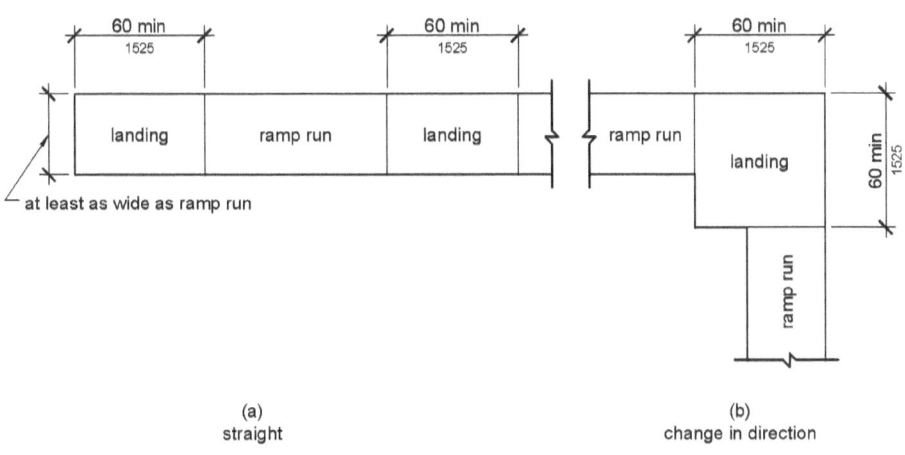

Figure 405.7
Ramp Landings

405.7.1 Slope. Landings shall comply with 302. Changes in level are not permitted.
  EXCEPTION: Slopes not steeper than 1:48 shall be permitted.

405.7.2 Width. The landing clear width shall be at least as wide as the widest *ramp* run leading to the landing.

405.7.3 Length. The landing clear length shall be 60 inches (1525 mm) long minimum.

405.7.4 Change in Direction. *Ramps* that change direction between runs at landings shall have a clear landing 60 inches (1525 mm) minimum by 60 inches (1525 mm) minimum.

405.7.5 Doorways. Where doorways are located adjacent to a *ramp* landing, maneuvering clearances required by 404.2.4 and 404.3.2 shall be permitted to overlap the required landing area.

405.8 Handrails. *Ramp* runs with a rise greater than 6 inches (150 mm) shall have handrails complying with 505.
  EXCEPTION: Within *employee work area*s, handrails shall not be required where *ramps* that are part of *common use circulation paths* are designed to permit the installation of handrails complying with 505. *Ramps* not subject to the exception to 405.5 shall be designed to maintain a 36 inch (915 mm) minimum clear width when handrails are installed.

405.9 Edge Protection. Edge protection complying with 405.9.1 or 405.9.2 shall be provided on each side of *ramp* runs and at each side of *ramp* landings.

EXCEPTIONS: 1. Edge protection shall not be required on *ramps* that are not required to have handrails and have sides complying with 406.3.
2. Edge protection shall not be required on the sides of *ramp* landings serving an adjoining *ramp* run or stairway.
3. Edge protection shall not be required on the sides of *ramp* landings having a vertical drop-off of ½ inch (13 mm) maximum within 10 inches (255 mm) horizontally of the minimum landing area specified in 405.7.

405.9.1 Extended Floor or Ground Surface. The floor or ground surface of the *ramp* run or landing shall extend 12 inches (305 mm) minimum beyond the inside face of a handrail complying with 505.

> **Advisory 405.9.1 Extended Floor or Ground Surface.** The extended surface prevents wheelchair casters and crutch tips from slipping off the ramp surface.

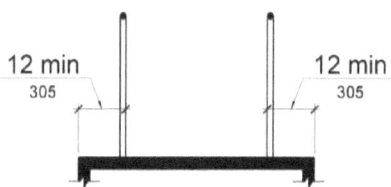

Figure 405.9.1
Extended Floor or Ground Surface Edge Protection

405.9.2 Curb or Barrier. A curb or barrier shall be provided that prevents the passage of a 4 inch (100 mm) diameter sphere, where any portion of the sphere is within 4 inches (100 mm) of the finish floor or ground surface.

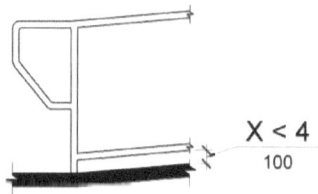

Figure 405.9.2
Curb or Barrier Edge Protection

405.10 Wet Conditions. Landings subject to wet conditions shall be designed to prevent the accumulation of water.

406 Curb Ramps

406.1 General. *Curb ramps* on *accessible* routes shall comply with 406, 405.2 through 405.5, and 405.10.

406.2 Counter Slope. Counter slopes of adjoining gutters and road surfaces immediately adjacent to the *curb ramp* shall not be steeper than 1:20. The adjacent surfaces at transitions at *curb ramps* to *walks*, gutters, and streets shall be at the same level.

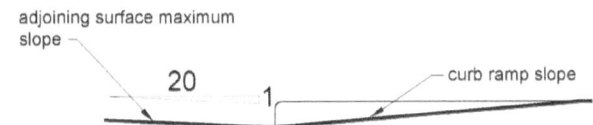

Figure 406.2
Counter Slope of Surfaces Adjacent to Curb Ramps

406.3 Sides of Curb Ramps. Where provided, *curb ramp* flares shall not be steeper than 1:10.

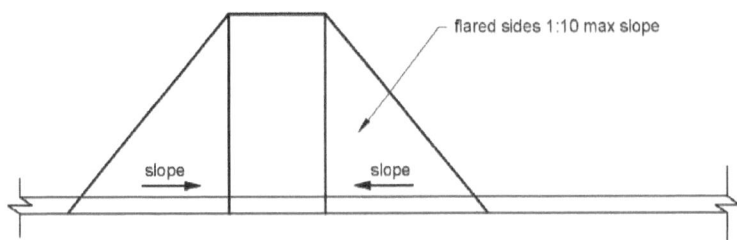

Figure 406.3
Sides of Curb Ramps

406.4 Landings. Landings shall be provided at the tops of *curb ramps*. The landing clear length shall be 36 inches (915 mm) minimum. The landing clear width shall be at least as wide as the *curb ramp*, excluding flared sides, leading to the landing.

EXCEPTION: In *alterations*, where there is no landing at the top of *curb ramps*, *curb ramp* flares shall be provided and shall not be steeper than 1:12.

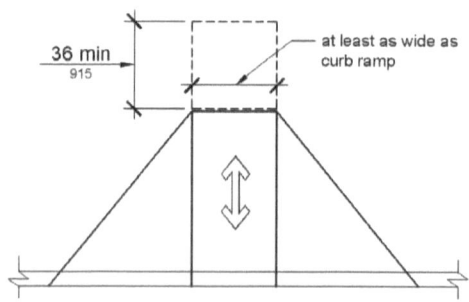

Figure 406.4
Landings at the Top of Curb Ramps

406.5 Location. *Curb ramps* and the flared sides of *curb ramps* shall be located so that they do not project into vehicular traffic lanes, parking *spaces*, or parking access aisles. *Curb ramps* at *marked crossings* shall be wholly contained within the markings, excluding any flared sides.

406.6 Diagonal Curb Ramps. Diagonal or corner type *curb ramps* with returned curbs or other well-defined edges shall have the edges parallel to the direction of pedestrian flow. The bottom of diagonal *curb ramps* shall have a clear *space* 48 inches (1220 mm) minimum outside active traffic lanes of the roadway. Diagonal *curb ramps* provided at *marked crossings* shall provide the 48 inches (1220 mm) minimum clear *space* within the markings. Diagonal *curb ramps* with flared sides shall have a segment of curb 24 inches (610 mm) long minimum located on each side of the *curb ramp* and within the *marked crossing*.

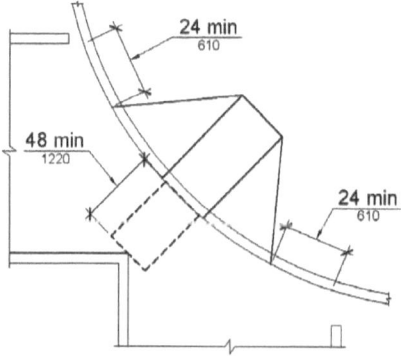

Figure 406.6
Diagonal or Corner Type Curb Ramps

406.7 Islands. Raised islands in crossings shall be cut through level with the street or have *curb ramps* at both sides. Each *curb ramp* shall have a level area 48 inches (1220 mm) long minimum by 36 inches (915 mm) wide minimum at the top of the *curb ramp* in the part of the island intersected by the crossings. Each 48 inch (1220 mm) minimum by 36 inch (915 mm) minimum area shall be oriented so that the 48 inch (1220 mm) minimum length is in the direction of the *running slope* of the *curb ramp* it serves. The 48 inch (1220 mm) minimum by 36 inch (915 mm) minimum areas and the *accessible* route shall be permitted to overlap.

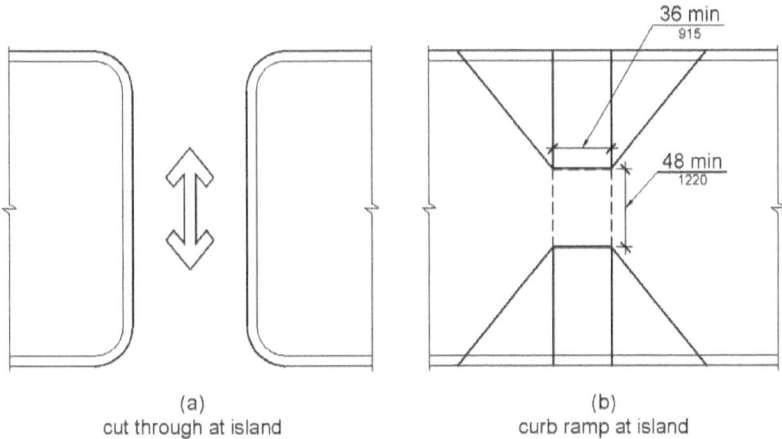

(a) cut through at island

(b) curb ramp at island

**Figure 406.7
Islands in Crossings**

407 Elevators

407.1 General. Elevators shall comply with 407 and with ASME A17.1 (incorporated by reference, see "Referenced Standards" in Chapter 1). They shall be passenger elevators as classified by ASME A17.1. Elevator operation shall be automatic.

> **Advisory 407.1 General.** The ADA and other Federal civil rights laws require that accessible features be maintained in working order so that they are accessible to and usable by those people they are intended to benefit. Building owners should note that the ASME Safety Code for Elevators and Escalators requires routine maintenance and inspections. Isolated or temporary interruptions in service due to maintenance or repairs may be unavoidable; however, failure to take prompt action to effect repairs could constitute a violation of Federal laws and these requirements.

407.2 Elevator Landing Requirements. Elevator landings shall comply with 407.2.

407.2.1 Call Controls. Where elevator call buttons or keypads are provided, they shall comply with 407.2.1 and 309.4. Call buttons shall be raised or flush.
   EXCEPTION: Existing elevators shall be permitted to have recessed call buttons.

407.2.1.1 Height. Call buttons and keypads shall be located within one of the reach ranges specified in 308, measured to the centerline of the highest *operable part*.
   EXCEPTION: Existing call buttons and existing keypads shall be permitted to be located at 54 inches (1370 mm) maximum above the finish floor, measured to the centerline of the highest *operable part*.

407.2.1.2 Size. Call buttons shall be ¾ inch (19 mm) minimum in the smallest dimension.
   EXCEPTION: Existing elevator call buttons shall not be required to comply with 407.2.1.2.

407.2.1.3 Clear Floor or Ground Space. A clear floor or ground *space* complying with 305 shall be provided at call controls.

> **Advisory 407.2.1.3 Clear Floor or Ground Space.** The clear floor or ground space required at elevator call buttons must remain free of obstructions including ashtrays, plants, and other decorative elements that prevent wheelchair users and others from reaching the call buttons. The height of the clear floor or ground space is considered to be a volume from the floor to 80 inches (2030 mm) above the floor. Recessed ashtrays should not be placed near elevator call buttons so that persons who are blind or visually impaired do not inadvertently contact them or their contents as they reach for the call buttons.

407.2.1.4 Location. The call button that designates the up direction shall be located above the call button that designates the down direction.
   EXCEPTION: Destination-oriented elevators shall not be required to comply with 407.2.1.4.

> **Advisory 407.2.1.4 Location Exception.** A destination-oriented elevator system provides lobby controls enabling passengers to select floor stops, lobby indicators designating which elevator to use, and a car indicator designating the floors at which the car will stop. Responding cars are programmed for maximum efficiency by reducing the number of stops any passenger experiences.

407.2.1.5 Signals. Call buttons shall have visible signals to indicate when each call is registered and when each call is answered.
   EXCEPTIONS: 1. Destination-oriented elevators shall not be required to comply with 407.2.1.5 provided that visible and audible signals complying with 407.2.2 indicating which elevator car to enter are provided.
   2. Existing elevators shall not be required to comply with 407.2.1.5.

407.2.1.6 Keypads. Where keypads are provided, keypads shall be in a standard telephone keypad arrangement and shall comply with 407.4.7.2.

407.2.2 Hall Signals. Hall signals, including in-car signals, shall comply with 407.2.2.

407.2.2.1 Visible and Audible Signals. A visible and audible signal shall be provided at each hoistway entrance to indicate which car is answering a call and the car's direction of travel. Where in-car signals are provided, they shall be visible from the floor area adjacent to the hall call buttons.

> EXCEPTIONS: 1. Visible and audible signals shall not be required at each destination-oriented elevator where a visible and audible signal complying with 407.2.2 is provided indicating the elevator car designation information.
> 2. In existing elevators, a signal indicating the direction of car travel shall not be required.

407.2.2.2 Visible Signals. Visible signal fixtures shall be centered at 72 inches (1830 mm) minimum above the finish floor or ground. The visible signal *elements* shall be 2-½ inches (64 mm) minimum measured along the vertical centerline of the *element*. Signals shall be visible from the floor area adjacent to the hall call button.

> EXCEPTIONS: 1. Destination-oriented elevators shall be permitted to have signals visible from the floor area adjacent to the hoistway entrance.
> 2. Existing elevators shall not be required to comply with 407.2.2.2.

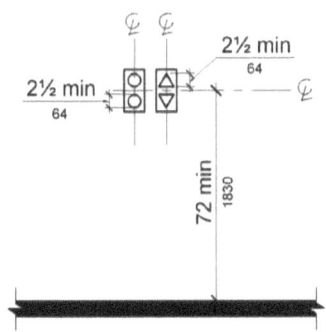

Figure 407.2.2.2
Visible Hall Signals

407.2.2.3 Audible Signals. Audible signals shall sound once for the up direction and twice for the down direction, or shall have verbal annunciators that indicate the direction of elevator car travel. Audible signals shall have a frequency of 1500 Hz maximum. Verbal annunciators shall have a frequency of 300 Hz minimum and 3000 Hz maximum. The audible signal and verbal annunciator shall be 10 dB minimum above ambient, but shall not exceed 80 dB, measured at the hall call button.

> EXCEPTIONS: 1. Destination-oriented elevators shall not be required to comply with 407.2.2.3 provided that the audible tone and verbal announcement is the same as those given at the call button or call button keypad.
> 2. Existing elevators shall not be required to comply with the requirements for frequency and dB range of audible signals.

407.2.2.4 Differentiation. Each destination-oriented elevator in a bank of elevators shall have audible and visible means for differentiation.

407.2.3 Hoistway Signs. Signs at elevator hoistways shall comply with 407.2.3.

407.2.3.1 Floor Designation. Floor designations complying with 703.2 and 703.4.1 shall be provided on both jambs of elevator hoistway entrances. Floor designations shall be provided in both *tactile characters* and braille. *Tactile characters* shall be 2 inches (51 mm) high minimum. A *tactile* star shall be provided on both jambs at the main entry level.

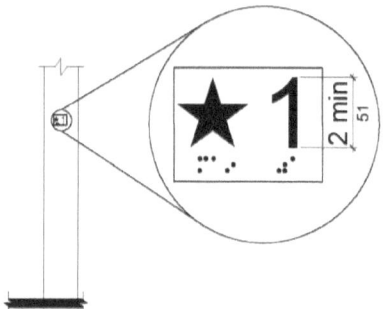

Figure 407.2.3.1
Floor Designations on Jambs of Elevator Hoistway Entrances

407.2.3.2 Car Designations. Destination-oriented elevators shall provide *tactile* car identification complying with 703.2 on both jambs of the hoistway immediately below the floor designation. Car designations shall be provided in both *tactile characters* and braille. *Tactile characters* shall be 2 inches (51 mm) high minimum.

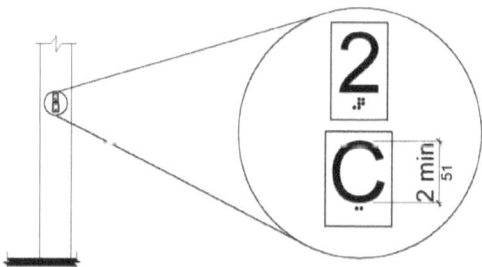

Figure 407.2.3.2
Car Designations on Jambs of Destination-Oriented Elevator Hoistway Entrances

407.3 Elevator Door Requirements. Hoistway and car doors shall comply with 407.3.

407.3.1 Type. Elevator doors shall be the horizontal sliding type. Car gates shall be prohibited.

407.3.2 Operation. Elevator hoistway and car doors shall open and close automatically.
EXCEPTION: Existing manually operated hoistway swing doors shall be permitted provided that they comply with 404.2.3 and 404.2.9. Car door closing shall not be initiated until the hoistway door is closed.

407.3.3 Reopening Device. Elevator doors shall be provided with a reopening device complying with 407.3.3 that shall stop and reopen a car door and hoistway door automatically if the door becomes obstructed by an object or person.
EXCEPTION: Existing elevators with manually operated doors shall not be required to comply with 407.3.3.

407.3.3.1 Height. The device shall be activated by sensing an obstruction passing through the opening at 5 inches (125 mm) nominal and 29 inches (735 mm) nominal above the finish floor.

407.3.3.2 Contact. The device shall not require physical contact to be activated, although contact is permitted to occur before the door reverses.

407.3.3.3 Duration. Door reopening devices shall remain effective for 20 seconds minimum.

407.3.4 Door and Signal Timing. The minimum acceptable time from notification that a car is answering a call or notification of the car assigned at the means for the entry of destination information until the doors of that car start to close shall be calculated from the following equation:

$T = D/(1.5 \text{ ft/s})$ or $T = D/(455 \text{ mm/s}) = 5$ seconds minimum where T equals the total time in seconds and D equals the distance (in feet or millimeters) from the point in the lobby or corridor 60 inches (1525 mm) directly in front of the farthest call button controlling that car to the centerline of its hoistway door.
EXCEPTIONS: 1. For cars with in-car lanterns, T shall be permitted to begin when the signal is visible from the point 60 inches (1525 mm) directly in front of the farthest hall call button and the audible signal is sounded.
2. Destination-oriented elevators shall not be required to comply with 407.3.4.

407.3.5 Door Delay. Elevator doors shall remain fully open in response to a car call for 3 seconds minimum.

407.3.6 Width. The width of elevator doors shall comply with Table 407.4.1.
EXCEPTION: In existing elevators, a power-operated car door complying with 404.2.3 shall be permitted.

407.4 Elevator Car Requirements. Elevator cars shall comply with 407.4.

407.4.1 Car Dimensions. Inside dimensions of elevator cars and clear width of elevator doors shall comply with Table 407.4.1.

EXCEPTION: Existing elevator car configurations that provide a clear floor area of 16 square feet (1.5 m²) minimum and also provide an inside clear depth 54 inches (1370 mm) minimum and a clear width 36 inches (915 mm) minimum shall be permitted.

Table 407.4.1 Elevator Car Dimensions

Door Location	Door Clear Width	Minimum Dimensions		
		Inside Car, Side to Side	Inside Car, Back Wall to Front Return	Inside Car, Back Wall to Inside Face of Door
Centered	42 inches (1065 mm)	80 inches (2030 mm)	51 inches (1295 mm)	54 inches (1370 mm)
Side (off-centered)	36 inches (915 mm)[1]	68 inches (1725 mm)	51 inches (1295 mm)	54 inches (1370 mm)
Any	36 inches (915 mm)[1]	54 inches (1370 mm)	80 inches (2030 mm)	80 inches (2030 mm)
Any	36 inches (915 mm)[1]	60 inches (1525 mm)[2]	60 inches (1525 mm)[2]	60 inches (1525 mm)[2]

1. A tolerance of minus 5/8 inch (16 mm) is permitted.
2. Other car configurations that provide a turning *space* complying with 304 with the door closed shall be permitted.

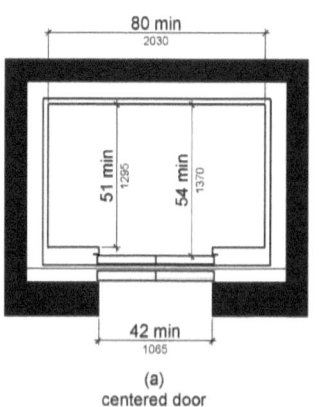

(a) centered door

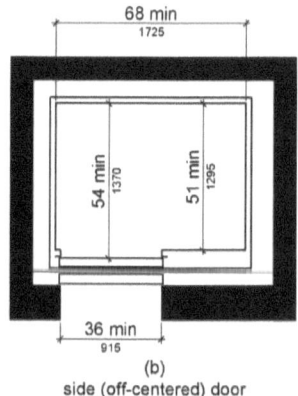

(b) side (off-centered) door

Figure 407.4.1
Elevator Car Dimensions

CHAPTER 4: ACCESSIBLE ROUTES   TECHNICAL

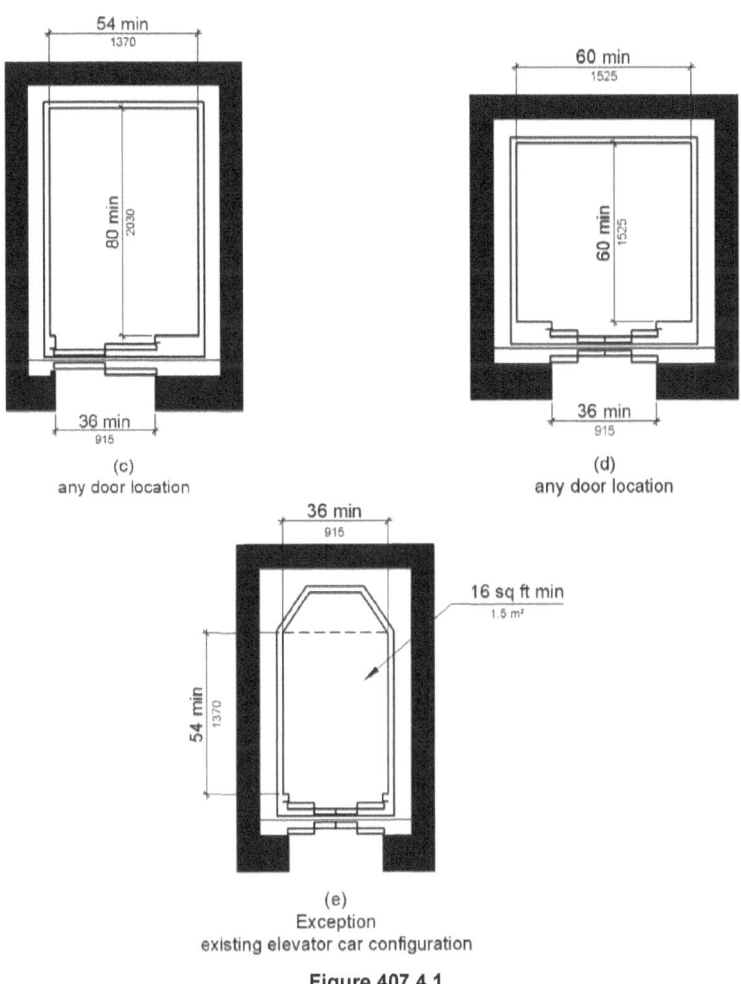

Figure 407.4.1
Elevator Car Dimensions

407.4.2 Floor Surfaces. Floor surfaces in elevator cars shall comply with 302 and 303.

407.4.3 Platform to Hoistway Clearance. The clearance between the car platform sill and the edge of any hoistway landing shall be 1¼ inch (32 mm) maximum.

407.4.4 Leveling. Each car shall be equipped with a self-leveling feature that will automatically bring and maintain the car at floor landings within a tolerance of ½ inch (13 mm) under rated loading to zero loading conditions.

407.4.5 Illumination. The level of illumination at the car controls, platform, car threshold and car landing sill shall be 5 foot candles (54 lux) minimum.

407.4.6 Elevator Car Controls. Where provided, elevator car controls shall comply with 407.4.6 and 309.4.
> **EXCEPTION:** In existing elevators, where a new car operating panel complying with 407.4.6 is provided, existing car operating panels shall not be required to comply with 407.4.6.

> **407.4.6.1 Location.** Controls shall be located within one of the reach ranges specified in 308.
> **EXCEPTIONS: 1.** Where the elevator panel serves more than 16 openings and a parallel approach is provided, buttons with floor designations shall be permitted to be 54 inches (1370 mm) maximum above the finish floor.
> **2.** In existing elevators, car control buttons with floor designations shall be permitted to be located 54 inches (1370 mm) maximum above the finish floor where a parallel approach is provided.

> **407.4.6.2 Buttons.** Car control buttons with floor designations shall comply with 407.4.6.2 and shall be raised or flush.
> **EXCEPTION:** In existing elevators, buttons shall be permitted to be recessed.

>> **407.4.6.2.1 Size.** Buttons shall be 3/4 inch (19 mm) minimum in their smallest dimension.

>> **407.4.6.2.2 Arrangement.** Buttons shall be arranged with numbers in ascending order. When two or more columns of buttons are provided they shall read from left to right.

> **407.4.6.3 Keypads.** Car control keypads shall be in a standard telephone keypad arrangement and shall comply with 407.4.7.2.

> **407.4.6.4 Emergency Controls.** Emergency controls shall comply with 407.4.6.4.

>> **407.4.6.4.1 Height.** Emergency control buttons shall have their centerlines 35 inches (890 mm) minimum above the finish floor.

>> **407.4.6.4.2 Location.** Emergency controls, including the emergency alarm, shall be grouped at the bottom of the panel.

407.4.7 Designations and Indicators of Car Controls. Designations and indicators of car controls shall comply with 407.4.7.

CHAPTER 4: ACCESSIBLE ROUTES   TECHNICAL

EXCEPTION: In existing elevators, where a new car operating panel complying with 407.4.7 is provided, existing car operating panels shall not be required to comply with 407.4.7.

407.4.7.1 Buttons. Car control buttons shall comply with 407.4.7.1.

407.4.7.1.1 Type. Control buttons shall be identified by *tactile characters* complying with 703.2.

407.4.7.1.2 Location. Raised *character* and braille designations shall be placed immediately to the left of the control button to which the designations apply.
    EXCEPTION: Where *space* on an existing car operating panel precludes *tactile* markings to the left of the controls, markings shall be placed as near to the control as possible.

407.4.7.1.3 Symbols. The control button for the emergency stop, alarm, door open, door close, main entry floor, and phone, shall be identified with *tactile* symbols as shown in Table 407.4.7.1.3.

Table 407.4.7.1.3 Elevator Control Button Identification

Control Button	Tactile Symbol	Braille Message
Emergency Stop	⊗	"ST"OP Three cells
Alarm	🔔	AL"AR"M Four cells
Door Open	◀║▶	OP"EN" Three cells
Door Close	▶║◀	CLOSE Five cells
Main Entry Floor	★	MA"IN" Three cells
Phone	☎	PH"ONE" Four cells

407.4.7.1.4 Visible Indicators. Buttons with floor designations shall be provided with visible indicators to show that a call has been registered. The visible indication shall extinguish when the car arrives at the designated floor.

407.4.7.2 Keypads. Keypads shall be identified by *characters* complying with 703.5 and shall be centered on the corresponding keypad button. The number five key shall have a single raised dot. The dot shall be 0.118 inch (3 mm) to 0.120 inch (3.05 mm) base diameter and in other aspects comply with Table 703.3.1.

407.4.8 Car Position Indicators. Audible and visible car position indicators shall be provided in elevator cars.

407.4.8.1 Visible Indicators. Visible indicators shall comply with 407.4.8.1.

407.4.8.1.1 Size. *Characters* shall be ½ inch (13 mm) high minimum.

407.4.8.1.2 Location. Indicators shall be located above the car control panel or above the door.

407.4.8.1.3 Floor Arrival. As the car passes a floor and when a car stops at a floor served by the elevator, the corresponding *character* shall illuminate.
EXCEPTION: Destination-oriented elevators shall not be required to comply with 407.4.8.1.3 provided that the visible indicators extinguish when the call has been answered.

407.4.8.1.4 Destination Indicator. In destination-oriented elevators, a display shall be provided in the car with visible indicators to show car destinations.

407.4.8.2 Audible Indicators. Audible indicators shall comply with 407.4.8.2.

407.4.8.2.1 Signal Type. The signal shall be an automatic verbal annunciator which announces the floor at which the car is about to stop.
EXCEPTION: For elevators other than destination-oriented elevators that have a rated speed of 200 feet per minute (1 m/s) or less, a non-verbal audible signal with a frequency of 1500 Hz maximum which sounds as the car passes or is about to stop at a floor served by the elevator shall be permitted.

407.4.8.2.2 Signal Level. The verbal annunciator shall be 10 dB minimum above ambient, but shall not exceed 80 dB, measured at the annunciator.

407.4.8.2.3 Frequency. The verbal annunciator shall have a frequency of 300 Hz minimum to 3000 Hz maximum.

407.4.9 Emergency Communication. Emergency two-way communication systems shall comply with 308. *Tactile* symbols and *characters* shall be provided adjacent to the device and shall comply with 703.2.

408 Limited-Use/Limited-Application Elevators

408.1 General. Limited-use/limited-application elevators shall comply with 408 and with ASME A17.1 (incorporated by reference, see "Referenced Standards" in Chapter 1). They shall be passenger elevators as classified by ASME A17.1. Elevator operation shall be automatic.

408.2 Elevator Landings. Landings serving limited-use/limited-application elevators shall comply with 408.2.

408.2.1 Call Buttons. Elevator call buttons and keypads shall comply with 407.2.1.

408.2.2 Hall Signals. Hall signals shall comply with 407.2.2.

408.2.3 Hoistway Signs. Signs at elevator hoistways shall comply with 407.2.3.1.

408.3 Elevator Doors. Elevator hoistway doors shall comply with 408.3.

408.3.1 Sliding Doors. Sliding hoistway and car doors shall comply with 407.3.1 through 407.3.3 and 408.4.1.

408.3.2 Swinging Doors. Swinging hoistway doors shall open and close automatically and shall comply with 404, 407.3.2 and 408.3.2.

408.3.2.1 Power Operation. Swinging doors shall be power-operated and shall comply with ANSI/BHMA A156.19 (1997 or 2002 edition) (incorporated by reference, see "Referenced Standards" in Chapter 1).

408.3.2.2 Duration. Power-operated swinging doors shall remain open for 20 seconds minimum when activated.

408.4 Elevator Cars. Elevator cars shall comply with 408.4.

408.4.1 Car Dimensions and Doors. Elevator cars shall provide a clear width 42 inches (1065 mm) minimum and a clear depth 54 inches (1370 mm) minimum. Car doors shall be positioned at the narrow ends of cars and shall provide 32 inches (815 mm) minimum clear width.
EXCEPTIONS: 1. Cars that provide a clear width 51 inches (1295 mm) minimum shall be permitted to provide a clear depth 51 inches (1295 mm) minimum provided that car doors provide a clear opening 36 inches (915 mm) wide minimum.
2. Existing elevator cars shall be permitted to provide a clear width 36 inches (915 mm) minimum, clear depth 54 inches (1370 mm) minimum, and a net clear platform area 15 square feet (1.4 m^2) minimum.

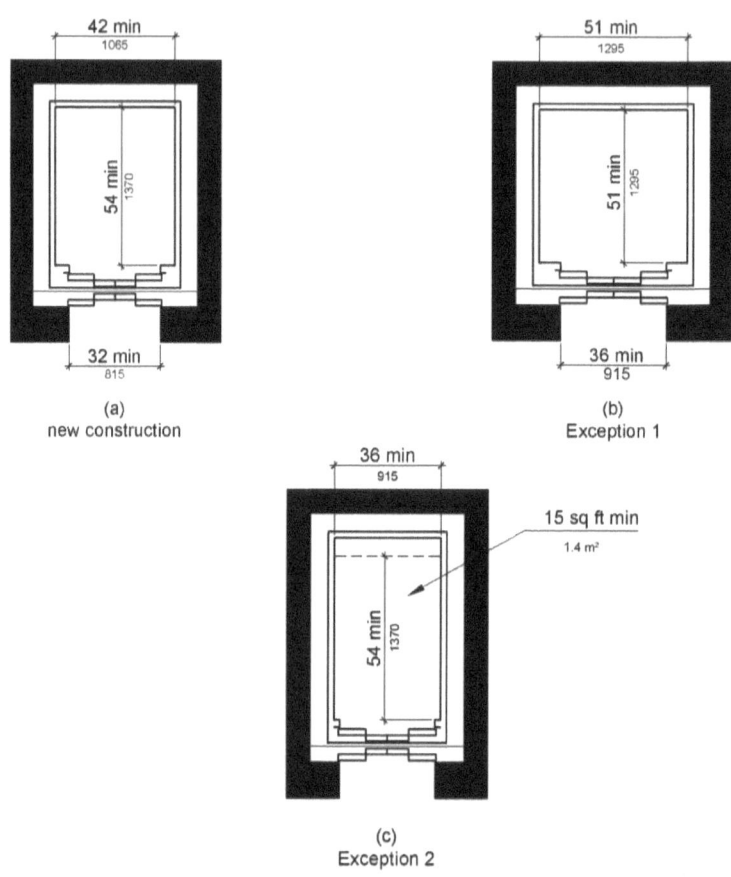

Figure 408.4.1
Limited-Use/Limited-Application (LULA) Elevator Car Dimensions

408.4.2 Floor Surfaces. Floor surfaces in elevator cars shall comply with 302 and 303.

408.4.3 Platform to Hoistway Clearance. The platform to hoistway clearance shall comply with 407.4.3.

408.4.4 Leveling. Elevator car leveling shall comply with 407.4.4.

CHAPTER 4: ACCESSIBLE ROUTES   TECHNICAL

408.4.5 Illumination. Elevator car illumination shall comply with 407.4.5.

408.4.6 Car Controls. Elevator car controls shall comply with 407.4.6. Control panels shall be centered on a side wall.

408.4.7 Designations and Indicators of Car Controls. Designations and indicators of car controls shall comply with 407.4.7.

408.4.8 Emergency Communications. Car emergency signaling devices complying with 407.4.9 shall be provided.

409 Private Residence Elevators

409.1 General. Private residence elevators that are provided within a *residential dwelling unit* required to provide mobility features complying with 809.2 through 809.4 shall comply with 409 and with ASME A17.1 (incorporated by reference, see "Referenced Standards" in Chapter 1). They shall be passenger elevators as classified by ASME A17.1. Elevator operation shall be automatic.

409.2 Call Buttons. Call buttons shall be ¾ inch (19 mm) minimum in the smallest dimension and shall comply with 309.

409.3 Elevator Doors. Hoistway doors, car doors, and car gates shall comply with 409.3 and 404.
   EXCEPTION: Doors shall not be required to comply with the maneuvering clearance requirements in 404.2.4.1 for approaches to the push side of swinging doors.

   409.3.1 Power Operation. Elevator car and hoistway doors and gates shall be power operated and shall comply with ANSI/BHMA A156.19 (1997 or 2002 edition) (incorporated by reference, see "Referenced Standards" in Chapter 1). Power operated doors and gates shall remain open for 20 seconds minimum when activated.
      EXCEPTION: In elevator cars with more than one opening, hoistway doors and gates shall be permitted to be of the manual-open, self-close type.

   409.3.2 Location. Elevator car doors or gates shall be positioned at the narrow end of the clear floor *spaces* required by 409.4.1.

409.4 Elevator Cars. Private residence elevator cars shall comply with 409.4.

   409.4.1 Inside Dimensions of Elevator Cars. Elevator cars shall provide a clear floor *space* of 36 inches (915 mm) minimum by 48 inches (1220 mm) minimum and shall comply with 305.

   409.4.2 Floor Surfaces. Floor surfaces in elevator cars shall comply with 302 and 303.

   409.4.3 Platform to Hoistway Clearance. The clearance between the car platform and the edge of any landing sill shall be 1½ inch (38 mm) maximum.

   409.4.4 Leveling. Each car shall automatically stop at a floor landing within a tolerance of ½ inch (13 mm) under rated loading to zero loading conditions.

409.4.5 Illumination Levels. Elevator car illumination shall comply with 407.4.5.

409.4.6 Car Controls. Elevator car control buttons shall comply with 409.4.6, 309.3, 309.4, and shall be raised or flush.

409.4.6.1 Size. Control buttons shall be 3/4 inch (19 mm) minimum in their smallest dimension.

409.4.6.2 Location. Control panels shall be on a side wall, 12 inches (305 mm) minimum from any adjacent wall.

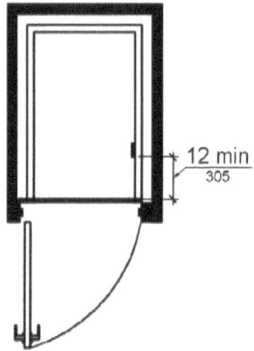

Figure 409.4.6.2
Location of Private Residence Elevator Control Panel

409.4.7 Emergency Communications. Emergency two-way communication systems shall comply with 409.4.7.

409.4.7.1 Type. A telephone and emergency signal device shall be provided in the car.

409.4.7.2 Operable Parts. The telephone and emergency signaling device shall comply with 309.3 and 309.4.

409.4.7.3 Compartment. If the telephone or device is in a closed compartment, the compartment door hardware shall comply with 309.

409.4.7.4 Cord. The telephone cord shall be 29 inches (735 mm) long minimum.

CHAPTER 4: ACCESSIBLE ROUTES — TECHNICAL

410 Platform Lifts

410.1 General. Platform lifts shall comply with ASME A18.1 (1999 edition or 2003 edition) (incorporated by reference, see "Referenced Standards" in Chapter 1). Platform lifts shall not be attendant-operated and shall provide unassisted entry and exit from the lift.

> **Advisory 410.1 General.** Inclined stairway chairlifts and inclined and vertical platform lifts are available for short-distance vertical transportation. Because an accessible route requires an 80 inch (2030 mm) vertical clearance, care should be taken in selecting lifts as they may not be equally suitable for use by people using wheelchairs and people standing. If a lift does not provide 80 inch (2030 mm) vertical clearance, it cannot be considered part of an accessible route in new construction.
>
> The ADA and other Federal civil rights laws require that accessible features be maintained in working order so that they are accessible to and usable by those people they are intended to benefit. Building owners are reminded that the ASME A18 Safety Standard for Platform Lifts and Stairway Chairlifts requires routine maintenance and inspections. Isolated or temporary interruptions in service due to maintenance or repairs may be unavoidable; however, failure to take prompt action to effect repairs could constitute a violation of Federal laws and these requirements.

410.2 Floor Surfaces. Floor surfaces in platform lifts shall comply with 302 and 303.

410.3 Clear Floor Space. Clear floor *space* in platform lifts shall comply with 305.

410.4 Platform to Runway Clearance. The clearance between the platform sill and the edge of any runway landing shall be 1¼ inch (32 mm) maximum.

410.5 Operable Parts. Controls for platform lifts shall comply with 309.

410.6 Doors and Gates. Platform lifts shall have low-energy power-operated doors or gates complying with 404.3. Doors shall remain open for 20 seconds minimum. End doors and gates shall provide a clear width 32 inches (815 mm) minimum. Side doors and gates shall provide a clear width 42 inches (1065 mm) minimum.
   EXCEPTION: Platform lifts serving two landings maximum and having doors or gates on opposite sides shall be permitted to have self-closing manual doors or gates.

TECHNICAL CHAPTER 4: ACCESSIBLE ROUTES

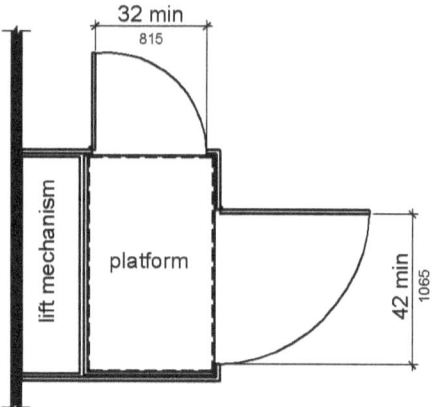

Figure 410.6
Platform Lift Doors and Gates

CHAPTER 5: GENERAL SITE AND BUILDING ELEMENTS

501 General

501.1 Scope. The provisions of Chapter 5 shall apply where required by Chapter 2 or where referenced by a requirement in this document.

502 Parking Spaces

502.1 General. Car and van parking *spaces* shall comply with 502. Where parking *spaces* are marked with lines, width measurements of parking *spaces* and access aisles shall be made from the centerline of the markings.
   EXCEPTION: Where parking *spaces* or access aisles are not adjacent to another parking *space* or access aisle, measurements shall be permitted to include the full width of the line defining the parking *space* or access aisle.

502.2 Vehicle Spaces. Car parking *spaces* shall be 96 inches (2440 mm) wide minimum and van parking *spaces* shall be 132 inches (3350 mm) wide minimum, shall be marked to define the width, and shall have an adjacent access aisle complying with 502.3.
   EXCEPTION: Van parking *spaces* shall be permitted to be 96 inches (2440 mm) wide minimum where the access aisle is 96 inches (2440 mm) wide minimum.

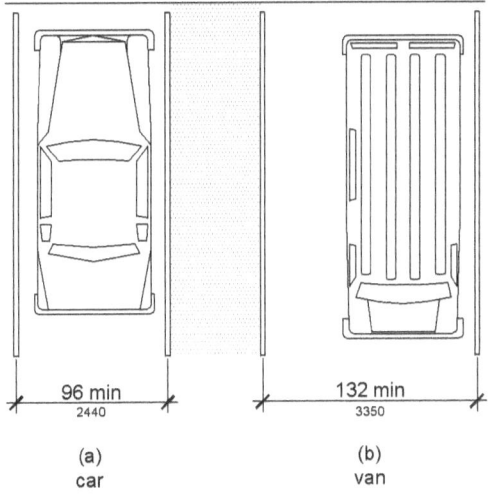

Figure 502.2
Vehicle Parking Spaces

502.3 Access Aisle. Access aisles serving parking *spaces* shall comply with 502.3. Access aisles shall adjoin an *accessible* route. Two parking *spaces* shall be permitted to share a common access aisle.

> **Advisory 502.3 Access Aisle.** Accessible routes must connect parking spaces to accessible entrances. In parking facilities where the accessible route must cross vehicular traffic lanes, marked crossings enhance pedestrian safety, particularly for people using wheelchairs and other mobility aids. Where possible, it is preferable that the accessible route not pass behind parked vehicles.

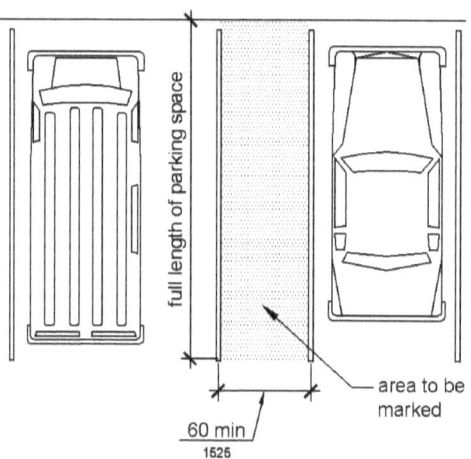

**Figure 502.3
Parking Space Access Aisle**

502.3.1 Width. Access aisles serving car and van parking *spaces* shall be 60 inches (1525 mm) wide minimum.

502.3.2 Length. Access aisles shall extend the full length of the parking *spaces* they serve.

502.3.3 Marking. Access aisles shall be marked so as to discourage parking in them.

> **Advisory 502.3.3 Marking.** The method and color of marking are not specified by these requirements but may be addressed by State or local laws or regulations. Because these requirements permit the van access aisle to be as wide as a parking space, it is important that the aisle be clearly marked.

502.3.4 Location. Access aisles shall not overlap the *vehicular way*. Access aisles shall be permitted to be placed on either side of the parking *space* except for angled van parking *spaces* which shall have access aisles located on the passenger side of the parking *spaces*.

> **Advisory 502.3.4 Location.** Wheelchair lifts typically are installed on the passenger side of vans. Many drivers, especially those who operate vans, find it more difficult to back into parking spaces than to back out into comparatively unrestricted vehicular lanes. For this reason, where a van and car share an access aisle, consider locating the van space so that the access aisle is on the passenger side of the van space.

502.4 Floor or Ground Surfaces. Parking *spaces* and access aisles serving them shall comply with 302. Access aisles shall be at the same level as the parking *spaces* they serve. Changes in level are not permitted.

EXCEPTION: Slopes not steeper than 1:48 shall be permitted.

> **Advisory 502.4 Floor or Ground Surfaces.** Access aisles are required to be nearly level in all directions to provide a surface for wheelchair transfer to and from vehicles. The exception allows sufficient slope for drainage. Built-up curb ramps are not permitted to project into access aisles and parking spaces because they would create slopes greater than 1:48.

502.5 Vertical Clearance. Parking *spaces* for vans and access aisles and vehicular routes serving them shall provide a vertical clearance of 98 inches (2490 mm) minimum.

> **Advisory 502.5 Vertical Clearance.** Signs provided at entrances to parking facilities informing drivers of clearances and the location of van accessible parking spaces can provide useful customer assistance.

502.6 Identification. Parking *space* identification signs shall include the International Symbol of *Accessibility* complying with 703.7.2.1. Signs identifying van parking *spaces* shall contain the designation "van accessible." Signs shall be 60 inches (1525 mm) minimum above the finish floor or ground surface measured to the bottom of the sign.

> **Advisory 502.6 Identification.** The required "van accessible" designation is intended to be informative, not restrictive, in identifying those spaces that are better suited for van use. Enforcement of motor vehicle laws, including parking privileges, is a local matter.

502.7 Relationship to Accessible Routes. Parking *spaces* and access aisles shall be designed so that cars and vans, when parked, cannot obstruct the required clear width of adjacent *accessible* routes.

> **Advisory 502.7 Relationship to Accessible Routes.** Wheel stops are an effective way to prevent vehicle overhangs from reducing the clear width of accessible routes.

503 Passenger Loading Zones

503.1 General. Passenger loading zones shall comply with 503.

503.2 Vehicle Pull-Up Space. Passenger loading zones shall provide a vehicular pull-up *space* 96 inches (2440 mm) wide minimum and 20 feet (6100 mm) long minimum.

503.3 Access Aisle. Passenger loading zones shall provide access aisles complying with 503 adjacent to the vehicle pull-up space. Access aisles shall adjoin an *accessible* route and shall not overlap the *vehicular way*.

503.3.1 Width. Access aisles serving vehicle pull-up *spaces* shall be 60 inches (1525 mm) wide minimum.

503.3.2 Length. Access aisles shall extend the full length of the vehicle pull-up *spaces* they serve.

503.3.3 Marking. Access aisles shall be marked so as to discourage parking in them.

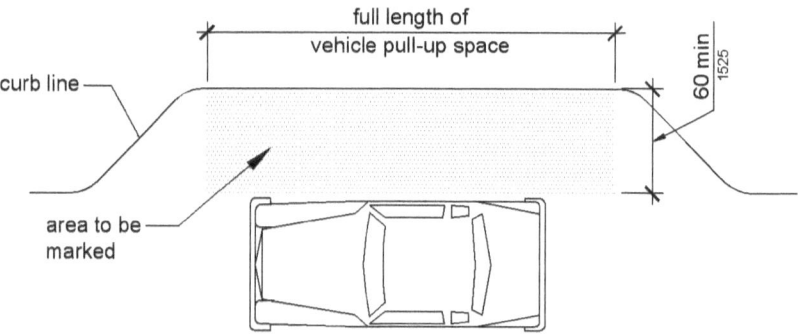

Figure 503.3
Passenger Loading Zone Access Aisle

503.4 Floor and Ground Surfaces. Vehicle pull-up *spaces* and access aisles serving them shall comply with 302. Access aisles shall be at the same level as the vehicle pull-up *space* they serve. Changes in level are not permitted.
   EXCEPTION: Slopes not steeper than 1:48 shall be permitted.

503.5 Vertical Clearance. Vehicle pull-up *spaces*, access aisles serving them, and a vehicular route from an *entrance* to the passenger loading zone, and from the passenger loading zone to a vehicular exit shall provide a vertical clearance of 114 inches (2895 mm) minimum.

504 Stairways

504.1 General. Stairs shall comply with 504.

504.2 Treads and Risers. All steps on a flight of stairs shall have uniform riser heights and uniform tread depths. Risers shall be 4 inches (100 mm) high minimum and 7 inches (180 mm) high maximum. Treads shall be 11 inches (280 mm) deep minimum.

504.3 Open Risers. Open risers are not permitted.

504.4 Tread Surface. Stair treads shall comply with 302. Changes in level are not permitted.
   EXCEPTION: Treads shall be permitted to have a slope not steeper than 1:48.

> **Advisory 504.4 Tread Surface.** Consider providing visual contrast on tread nosings, or at the leading edges of treads without nosings, so that stair treads are more visible for people with low vision.

504.5 Nosings. The radius of curvature at the leading edge of the tread shall be ½ inch (13 mm) maximum. Nosings that project beyond risers shall have the underside of the leading edge curved or beveled. Risers shall be permitted to slope under the tread at an angle of 30 degrees maximum from vertical. The permitted projection of the nosing shall extend 1½ inches (38 mm) maximum over the tread below.

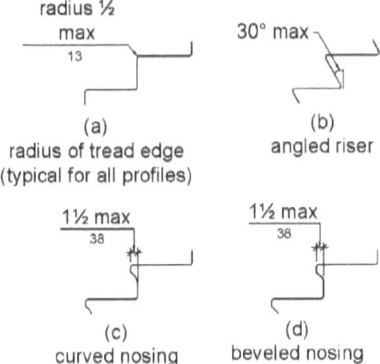

Figure 504.5
Stair Nosings

504.6 Handrails. Stairs shall have handrails complying with 505.

504.7 Wet Conditions. Stair treads and landings subject to wet conditions shall be designed to prevent the accumulation of water.

505 Handrails

505.1 General. Handrails provided along walking surfaces complying with 403, required at *ramps* complying with 405, and required at stairs complying with 504 shall comply with 505.

> **Advisory 505.1 General.** Handrails are required on ramp runs with a rise greater than 6 inches (150 mm) (see 405.8) and on certain stairways (see 504). Handrails are not required on walking surfaces with running slopes less than 1:20. However, handrails are required to comply with 505 when they are provided on walking surfaces with running slopes less than 1:20 (see 403.6). Sections 505.2, 505.3, and 505.10 do not apply to handrails provided on walking surfaces with running slopes less than 1:20 as these sections only reference requirements for ramps and stairs.

505.2 Where Required. Handrails shall be provided on both sides of stairs and *ramps*.
EXCEPTION: In *assembly areas*, handrails shall not be required on both sides of aisle *ramps* where a handrail is provided at either side or within the aisle width.

505.3 Continuity. Handrails shall be continuous within the full length of each stair flight or *ramp* run. Inside handrails on switchback or dogleg stairs and *ramps* shall be continuous between flights or runs.
EXCEPTION: In *assembly areas*, handrails on *ramps* shall not be required to be continuous in aisles serving seating.

505.4 Height. Top of gripping surfaces of handrails shall be 34 inches (865 mm) minimum and 38 inches (965 mm) maximum vertically above walking surfaces, stair nosings, and *ramp* surfaces. Handrails shall be at a consistent height above walking surfaces, stair nosings, and *ramp* surfaces.

> **Advisory 505.4 Height.** The requirements for stair and ramp handrails in this document are for adults. When children are the principal users in a building or facility (e.g., elementary schools), a second set of handrails at an appropriate height can assist them and aid in preventing accidents. A maximum height of 28 inches (710 mm) measured to the top of the gripping surface from the ramp surface or stair nosing is recommended for handrails designed for children. Sufficient vertical clearance between upper and lower handrails, 9 inches (230 mm) minimum, should be provided to help prevent entrapment.

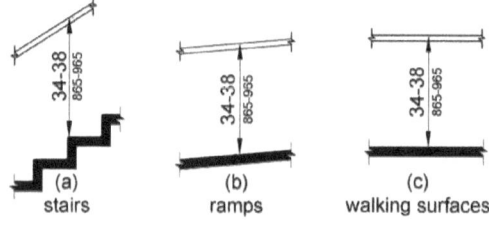

**Figure 505.4
Handrail Height**

505.5 Clearance. Clearance between handrail gripping surfaces and adjacent surfaces shall be 1½ inches (38 mm) minimum.

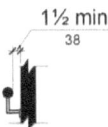

Figure 505.5
Handrail Clearance

505.6 Gripping Surface. Handrail gripping surfaces shall be continuous along their length and shall not be obstructed along their tops or sides. The bottoms of handrail gripping surfaces shall not be obstructed for more than 20 percent of their length. Where provided, horizontal projections shall occur 1½ inches (38 mm) minimum below the bottom of the handrail gripping surface.
> **EXCEPTIONS: 1.** Where handrails are provided along walking surfaces with slopes not steeper than 1:20, the bottoms of handrail gripping surfaces shall be permitted to be obstructed along their entire length where they are integral to crash rails or bumper guards.
> **2.** The distance between horizontal projections and the bottom of the gripping surface shall be permitted to be reduced by 1/8 inch (3.2 mm) for each ½ inch (13 mm) of additional handrail perimeter dimension that exceeds 4 inches (100 mm).

> **Advisory 505.6 Gripping Surface.** People with disabilities, older people, and others benefit from continuous gripping surfaces that permit users to reach the fingers outward or downward to grasp the handrail, particularly as the user senses a loss of equilibrium or begins to fall.

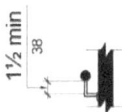

Figure 505.6
Horizontal Projections Below Gripping Surface

505.7 Cross Section. Handrail gripping surfaces shall have a cross section complying with 505.7.1 or 505.7.2.

505.7.1 Circular Cross Section. Handrail gripping surfaces with a circular cross section shall have an outside diameter of 1¼ inches (32 mm) minimum and 2 inches (51 mm) maximum.

505.7.2 Non-Circular Cross Sections. Handrail gripping surfaces with a non-circular cross section shall have a perimeter dimension of 4 inches (100 mm) minimum and 6¼ inches (160 mm) maximum, and a cross-section dimension of 2¼ inches (57 mm) maximum.

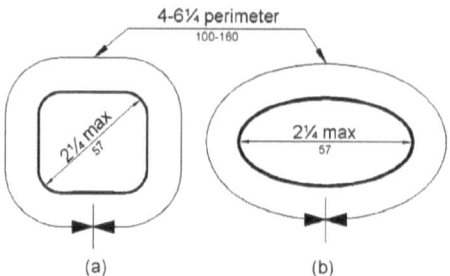

Figure 505.7.2
Handrail Non-Circular Cross Section

505.8 Surfaces. Handrail gripping surfaces and any surfaces adjacent to them shall be free of sharp or abrasive *elements* and shall have rounded edges.

505.9 Fittings. Handrails shall not rotate within their fittings.

505.10 Handrail Extensions. Handrail gripping surfaces shall extend beyond and in the same direction of stair flights and *ramp* runs in accordance with 505.10.
    EXCEPTIONS: 1. Extensions shall not be required for continuous handrails at the inside turn of switchback or dogleg stairs and *ramps*.
    2. In *assembly areas*, extensions shall not be required for *ramp* handrails in aisles serving seating where the handrails are discontinuous to provide access to seating and to permit crossovers within aisles.
    3. In *alterations*, full extensions of handrails shall not be required where such extensions would be hazardous due to plan configuration.

505.10.1 Top and Bottom Extension at Ramps. *Ramp* handrails shall extend horizontally above the landing for 12 inches (305 mm) minimum beyond the top and bottom of *ramp* runs. Extensions shall return to a wall, guard, or the landing surface, or shall be continuous to the handrail of an adjacent *ramp* run.

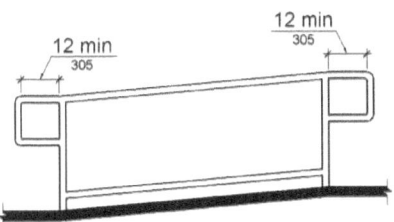

Figure 505.10.1
Top and Bottom Handrail Extension at Ramps

505.10.2 Top Extension at Stairs. At the top of a stair flight, handrails shall extend horizontally above the landing for 12 inches (305 mm) minimum beginning directly above the first riser nosing. Extensions shall return to a wall, guard, or the landing surface, or shall be continuous to the handrail of an adjacent stair flight.

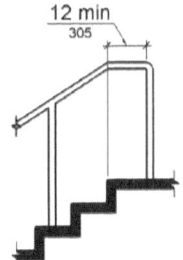

Figure 505.10.2
Top Handrail Extension at Stairs

505.10.3 Bottom Extension at Stairs. At the bottom of a stair flight, handrails shall extend at the slope of the stair flight for a horizontal distance at least equal to one tread depth beyond the last riser nosing. Extension shall return to a wall, guard, or the landing surface, or shall be continuous to the handrail of an adjacent stair flight.

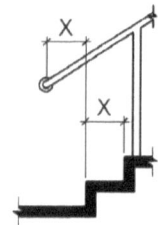

Note: X = tread depth

**Figure 505.10.3
Bottom Handrail Extension at Stairs**

CHAPTER 6: PLUMBING ELEMENTS AND FACILITIES

601 General

601.1 Scope. The provisions of Chapter 6 shall apply where required by Chapter 2 or where referenced by a requirement in this document.

602 Drinking Fountains

602.1 General. Drinking fountains shall comply with 307 and 602.

602.2 Clear Floor Space. Units shall have a clear floor or ground *space* complying with 305 positioned for a forward approach and centered on the unit. Knee and toe clearance complying with 306 shall be provided.
   EXCEPTION: A parallel approach complying with 305 shall be permitted at units for *children's use* where the spout is 30 inches (760 mm) maximum above the finish floor or ground and is 3½ inches (90 mm) maximum from the front edge of the unit, including bumpers.

602.3 Operable Parts. *Operable parts* shall comply with 309.

602.4 Spout Height. Spout outlets shall be 36 inches (915 mm) maximum above the finish floor or ground.

602.5 Spout Location. The spout shall be located 15 inches (380 mm) minimum from the vertical support and 5 inches (125 mm) maximum from the front edge of the unit, including bumpers.

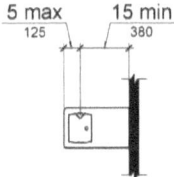

Figure 602.5
Drinking Fountain Spout Location

602.6 Water Flow. The spout shall provide a flow of water 4 inches (100 mm) high minimum and shall be located 5 inches (125 mm) maximum from the front of the unit. The angle of the water stream shall be measured horizontally relative to the front face of the unit. Where spouts are located less than 3 inches (75 mm) of the front of the unit, the angle of the water stream shall be 30 degrees maximum. Where spouts are located between 3 inches (75 mm) and 5 inches (125 mm) maximum from the front of the unit, the angle of the water stream shall be 15 degrees maximum.

> **Advisory 602.6 Water Flow.** The purpose of requiring the drinking fountain spout to produce a flow of water 4 inches (100 mm) high minimum is so that a cup can be inserted under the flow of water to provide a drink of water for an individual who, because of a disability, would otherwise be incapable of using the drinking fountain.

602.7 Drinking Fountains for Standing Persons. Spout outlets of drinking fountains for standing persons shall be 38 inches (965 mm) minimum and 43 inches (1090 mm) maximum above the finish floor or ground.

603 Toilet and Bathing Rooms

603.1 General. Toilet and bathing rooms shall comply with 603.

603.2 Clearances. Clearances shall comply with 603.2.

>**603.2.1 Turning Space.** Turning *space* complying with 304 shall be provided within the room.

>**603.2.2 Overlap.** Required clear floor *spaces*, clearance at fixtures, and turning *space* shall be permitted to overlap.

>**603.2.3 Door Swing.** Doors shall not swing into the clear floor *space* or clearance required for any fixture. Doors shall be permitted to swing into the required turning *space*.
>**EXCEPTIONS: 1.** Doors to a toilet room or bathing room for a single occupant accessed only through a private office and not for *common use* or *public use* shall be permitted to swing into the clear floor *space* or clearance provided the swing of the door can be reversed to comply with 603.2.3.
>**2.** Where the toilet room or bathing room is for individual use and a clear floor *space* complying with 305.3 is provided within the room beyond the arc of the door swing, doors shall be permitted to swing into the clear floor *space* or clearance required for any fixture.

> **Advisory 603.2.3 Door Swing Exception 1.** At the time the door is installed, and if the door swing is reversed in the future, the door must meet all the requirements specified in 404. Additionally, the door swing cannot reduce the required width of an accessible route. Also, avoid violating other building or life safety codes when the door swing is reversed.

603.3 Mirrors. Mirrors located above lavatories or countertops shall be installed with the bottom edge of the reflecting surface 40 inches (1015 mm) maximum above the finish floor or ground. Mirrors not located above lavatories or countertops shall be installed with the bottom edge of the reflecting surface 35 inches (890 mm) maximum above the finish floor or ground.

> **Advisory 603.3 Mirrors.** A single full-length mirror can accommodate a greater number of people, including children. In order for mirrors to be usable by people who are ambulatory and people who use wheelchairs, the top edge of mirrors should be 74 inches (1880 mm) minimum from the floor or ground.

603.4 Coat Hooks and Shelves. Coat hooks shall be located within one of the reach ranges specified in 308. Shelves shall be located 40 inches (1015 mm) minimum and 48 inches (1220 mm) maximum above the finish floor.

604 Water Closets and Toilet Compartments

604.1 General. Water closets and toilet compartments shall comply with 604.2 through 604.8.

EXCEPTION: Water closets and toilet compartments for *children's use* shall be permitted to comply with 604.9.

604.2 Location. The water closet shall be positioned with a wall or partition to the rear and to one side. The centerline of the water closet shall be 16 inches (405 mm) minimum to 18 inches (455 mm) maximum from the side wall or partition, except that the water closet shall be 17 inches (430 mm) minimum and 19 inches (485 mm) maximum from the side wall or partition in the ambulatory *accessible* toilet compartment specified in 604.8.2. Water closets shall be arranged for a left-hand or right-hand approach.

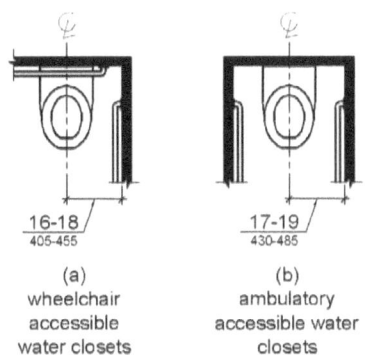

(a) wheelchair accessible water closets

(b) ambulatory accessible water closets

Figure 604.2
Water Closet Location

604.3 Clearance. Clearances around water closets and in toilet compartments shall comply with 604.3.

604.3.1 Size. Clearance around a water closet shall be 60 inches (1525 mm) minimum measured perpendicular from the side wall and 56 inches (1420 mm) minimum measured perpendicular from the rear wall.

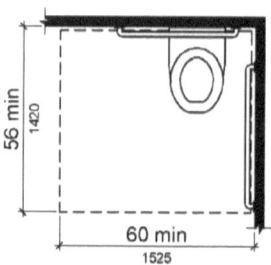

Figure 604.3.1
Size of Clearance at Water Closets

604.3.2 Overlap. The required clearance around the water closet shall be permitted to overlap the water closet, associated grab bars, dispensers, sanitary napkin disposal units, coat hooks, shelves, *accessible* routes, clear floor *space* and clearances required at other fixtures, and the turning *space*. No other fixtures or obstructions shall be located within the required water closet clearance.

> **EXCEPTION:** In *residential dwelling units*, a lavatory complying with 606 shall be permitted on the rear wall 18 inches (455 mm) minimum from the water closet centerline where the clearance at the water closet is 66 inches (1675 mm) minimum measured perpendicular from the rear wall.

Advisory 604.3.2 Overlap. When the door to the toilet room is placed directly in front of the water closet, the water closet cannot overlap the required maneuvering clearance for the door inside the room.

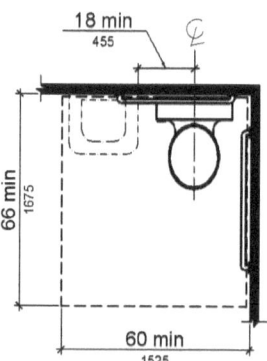

Figure 604.3.2 (Exception)
Overlap of Water Closet Clearance in Residential Dwelling Units

604.4 Seats. The seat height of a water closet above the finish floor shall be 17 inches (430 mm) minimum and 19 inches (485 mm) maximum measured to the top of the seat. Seats shall not be sprung to return to a lifted position.
    EXCEPTIONS: 1. A water closet in a toilet room for a single occupant accessed only through a private office and not for *common use* or *public use* shall not be required to comply with 604.4.
    2. In *residential dwelling units*, the height of water closets shall be permitted to be 15 inches (380 mm) minimum and 19 inches (485 mm) maximum above the finish floor measured to the top of the seat.

604.5 Grab Bars. Grab bars for water closets shall comply with 609. Grab bars shall be provided on the side wall closest to the water closet and on the rear wall.
    EXCEPTIONS: 1. Grab bars shall not be required to be installed in a toilet room for a single occupant accessed only through a private office and not for *common use* or *public use* provided that reinforcement has been installed in walls and located so as to permit the installation of grab bars complying with 604.5.
    2. In *residential dwelling units*, grab bars shall not be required to be installed in toilet or bathrooms provided that reinforcement has been installed in walls and located so as to permit the installation of grab bars complying with 604.5.
    3. In detention or correction *facilities*, grab bars shall not be required to be installed in housing or holding cells that are specially designed without protrusions for purposes of suicide prevention.

> **Advisory 604.5 Grab Bars Exception 2.** Reinforcement must be sufficient to permit the installation of rear and side wall grab bars that fully meet all accessibility requirements including, but not limited to, required length, installation height, and structural strength.

604.5.1 Side Wall. The side wall grab bar shall be 42 inches (1065 mm) long minimum, located 12 inches (305 mm) maximum from the rear wall and extending 54 inches (1370 mm) minimum from the rear wall.

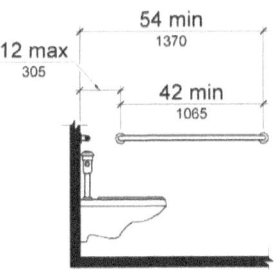

Figure 604.5.1
Side Wall Grab Bar at Water Closets

604.5.2 Rear Wall. The rear wall grab bar shall be 36 inches (915 mm) long minimum and extend from the centerline of the water closet 12 inches (305 mm) minimum on one side and 24 inches (610 mm) minimum on the other side.

EXCEPTIONS: 1. The rear grab bar shall be permitted to be 24 inches (610 mm) long minimum, centered on the water closet, where wall *space* does not permit a length of 36 inches (915 mm) minimum due to the location of a recessed fixture adjacent to the water closet.

2. Where an *administrative authority* requires flush controls for flush valves to be located in a position that conflicts with the location of the rear grab bar, then the rear grab bar shall be permitted to be split or shifted to the open side of the toilet area.

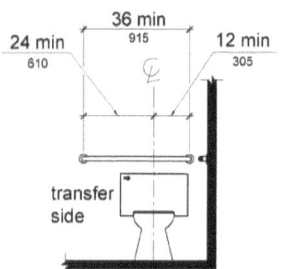

Figure 604.5.2
Rear Wall Grab Bar at Water Closets

604.6 Flush Controls. Flush controls shall be hand operated or automatic. Hand operated flush controls shall comply with 309. Flush controls shall be located on the open side of the water closet except in ambulatory *accessible* compartments complying with 604.8.2.

> **Advisory 604.6 Flush Controls.** If plumbing valves are located directly behind the toilet seat, flush valves and related plumbing can cause injury or imbalance when a person leans back against them. To prevent causing injury or imbalance, the plumbing can be located behind walls or to the side of the toilet; or if approved by the local authority having jurisdiction, provide a toilet seat lid.

604.7 Dispensers. Toilet paper dispensers shall comply with 309.4 and shall be 7 inches (180 mm) minimum and 9 inches (230 mm) maximum in front of the water closet measured to the centerline of the dispenser. The outlet of the dispenser shall be 15 inches (380 mm) minimum and 48 inches (1220 mm) maximum above the finish floor and shall not be located behind grab bars. Dispensers shall not be of a type that controls delivery or that does not allow continuous paper flow.

> **Advisory 604.7 Dispensers.** If toilet paper dispensers are installed above the side wall grab bar, the outlet of the toilet paper dispenser must be 48 inches (1220 mm) maximum above the finish floor and the top of the gripping surface of the grab bar must be 33 inches (840 mm) minimum and 36 inches (915 mm) maximum above the finish floor.

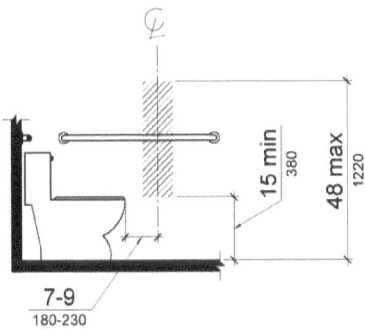

Figure 604.7
Dispenser Outlet Location

604.8 Toilet Compartments. Wheelchair *accessible* toilet compartments shall meet the requirements of 604.8.1 and 604.8.3. Compartments containing more than one plumbing fixture shall comply with 603. Ambulatory *accessible* compartments shall comply with 604.8.2 and 604.8.3.

604.8.1 Wheelchair Accessible Compartments. Wheelchair *accessible* compartments shall comply with 604.8.1.

604.8.1.1 Size. Wheelchair *accessible* compartments shall be 60 inches (1525 mm) wide minimum measured perpendicular to the side wall, and 56 inches (1420 mm) deep minimum for wall hung water closets and 59 inches (1500 mm) deep minimum for floor mounted water closets measured perpendicular to the rear wall. Wheelchair *accessible* compartments for *children's use* shall be 60 inches (1525 mm) wide minimum measured perpendicular to the side wall, and 59 inches (1500 mm) deep minimum for wall hung and floor mounted water closets measured perpendicular to the rear wall.

> **Advisory 604.8.1.1 Size.** The minimum space required in toilet compartments is provided so that a person using a wheelchair can maneuver into position at the water closet. This space cannot be obstructed by baby changing tables or other fixtures or conveniences, except as specified at 604.3.2 (Overlap). If toilet compartments are to be used to house fixtures other than those associated with the water closet, they must be designed to exceed the minimum space requirements. Convenience fixtures such as baby changing tables must also be accessible to people with disabilities as well as to other users. Toilet compartments that are designed to meet, and not exceed, the minimum space requirements may not provide adequate space for maneuvering into position at a baby changing table.

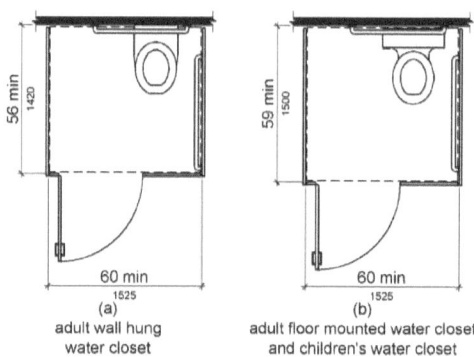

Figure 604.8.1.1
Size of Wheelchair Accessible Toilet Compartment

604.8.1.2 Doors. Toilet compartment doors, including door hardware, shall comply with 404 except that if the approach is to the latch side of the compartment door, clearance between the door side of the compartment and any obstruction shall be 42 inches (1065 mm) minimum. Doors shall be located in the front partition or in the side wall or partition farthest from the water closet. Where located in the front partition, the door opening shall be 4 inches (100 mm) maximum from the side wall or partition farthest from the water closet. Where located in the side wall or partition, the door opening shall be 4 inches (100 mm) maximum from the front partition. The door shall be self-closing. A door pull complying with 404.2.7 shall be placed on both sides of the door near the latch. Toilet compartment doors shall not swing into the minimum required compartment area.

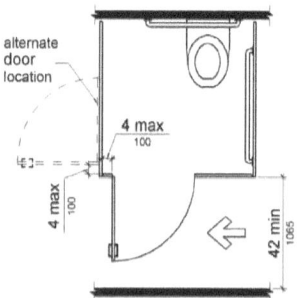

Figure 604.8.1.2
Wheelchair Accessible Toilet Compartment Doors

604.8.1.3 Approach. Compartments shall be arranged for left-hand or right-hand approach to the water closet.

604.8.1.4 Toe Clearance. The front partition and at least one side partition shall provide a toe clearance of 9 inches (230 mm) minimum above the finish floor and 6 inches (150 mm) deep minimum beyond the compartment-side face of the partition, exclusive of partition support members. Compartments for *children's use* shall provide a toe clearance of 12 inches (305 mm) minimum above the finish floor.

> **EXCEPTION:** Toe clearance at the front partition is not required in a compartment greater than 62 inches (1575 mm) deep with a wall-hung water closet or 65 inches (1650 mm) deep with a floor-mounted water closet. Toe clearance at the side partition is not required in a compartment greater than 66 inches (1675 mm) wide. Toe clearance at the front partition is not required in a compartment for *children's use* that is greater than 65 inches (1650 mm) deep.

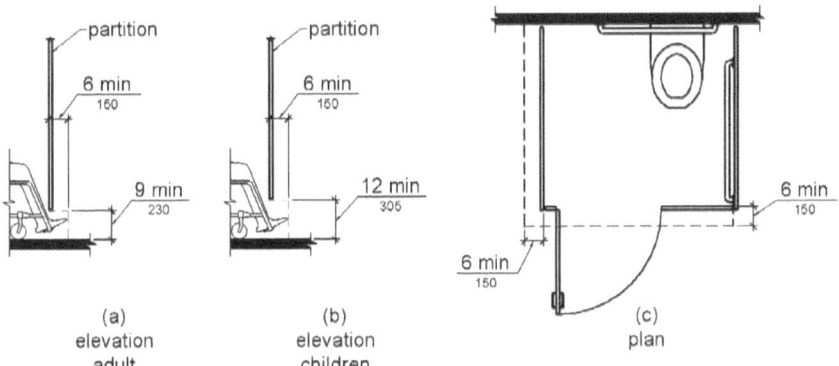

Figure 604.8.1.4
Wheelchair Accessible Toilet Compartment Toe Clearance

604.8.1.5 Grab Bars. Grab bars shall comply with 609. A side-wall grab bar complying with 604.5.1 shall be provided and shall be located on the wall closest to the water closet. In addition, a rear-wall grab bar complying with 604.5.2 shall be provided.

604.8.2 Ambulatory Accessible Compartments. Ambulatory *accessible* compartments shall comply with 604.8.2.

604.8.2.1 Size. Ambulatory *accessible* compartments shall have a depth of 60 inches (1525 mm) minimum and a width of 35 inches (890 mm) minimum and 37 inches (940 mm) maximum.

TECHNICAL   CHAPTER 6: PLUMBING ELEMENTS AND FACILITIES

604.8.2.2 Doors. Toilet compartment doors, including door hardware, shall comply with 404, except that if the approach is to the latch side of the compartment door, clearance between the door side of the compartment and any obstruction shall be 42 inches (1065 mm) minimum. The door shall be self-closing. A door pull complying with 404.2.7 shall be placed on both sides of the door near the latch. Toilet compartment doors shall not swing into the minimum required compartment area.

604.8.2.3 Grab Bars. Grab bars shall comply with 609. A side-wall grab bar complying with 604.5.1 shall be provided on both sides of the compartment.

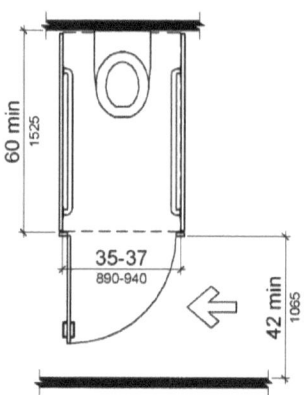

Figure 604.8.2
Ambulatory Accessible Toilet Compartment

604.8.3 Coat Hooks and Shelves. Coat hooks shall be located within one of the reach ranges specified in 308. Shelves shall be located 40 inches (1015 mm) minimum and 48 inches (1220 mm) maximum above the finish floor.

604.9 Water Closets and Toilet Compartments for Children's Use. Water closets and toilet compartments for *children's use* shall comply with 604.9.

> **Advisory 604.9 Water Closets and Toilet Compartments for Children's Use.** The requirements in 604.9 are to be followed where the exception for children's water closets in 604.1 is used. The following table provides additional guidance in applying the specifications for water closets for children according to the age group served and reflects the differences in the size, stature, and reach ranges of children ages 3 through 12. The specifications chosen should correspond to the age of the primary user group. The specifications of one age group should be applied consistently in the installation of a water closet and related elements.

Advisory Specifications for Water Closets Serving Children Ages 3 through 12			
	Ages 3 and 4	Ages 5 through 8	Ages 9 through 12
Water Closet Centerline	12 inches (305 mm)	12 to 15 inches (305 to 380 mm)	15 to 18 inches (380 to 455 mm)
Toilet Seat Height	11 to 12 inches (280 to 305 mm)	12 to 15 inches (305 to 380 mm)	15 to 17 inches (380 to 430 mm)
Grab Bar Height	18 to 20 inches (455 to 510 mm)	20 to 25 inches (510 to 635 mm)	25 to 27 inches (635 to 685 mm)
Dispenser Height	14 inches (355 mm)	14 to 17 inches (355 to 430 mm)	17 to 19 inches (430 to 485 mm)

604.9.1 Location. The water closet shall be located with a wall or partition to the rear and to one side. The centerline of the water closet shall be 12 inches (305 mm) minimum and 18 inches (455 mm) maximum from the side wall or partition, except that the water closet shall be 17 inches (430 mm) minimum and 19 inches (485 mm) maximum from the side wall or partition in the ambulatory *accessible* toilet compartment specified in 604.8.2. Compartments shall be arranged for left-hand or right-hand approach to the water closet.

604.9.2 Clearance. Clearance around a water closet shall comply with 604.3.

604.9.3 Height. The height of water closets shall be 11 inches (280 mm) minimum and 17 inches (430 mm) maximum measured to the top of the seat. Seats shall not be sprung to return to a lifted position.

604.9.4 Grab Bars. Grab bars for water closets shall comply with 604.5.

604.9.5 Flush Controls. Flush controls shall be hand operated or automatic. Hand operated flush controls shall comply with 309.2 and 309.4 and shall be installed 36 inches (915 mm) maximum above the finish floor. Flush controls shall be located on the open side of the water closet except in ambulatory *accessible* compartments complying with 604.8.2.

604.9.6 Dispensers. Toilet paper dispensers shall comply with 309.4 and shall be 7 inches (180 mm) minimum and 9 inches (230 mm) maximum in front of the water closet measured to the centerline of the dispenser. The outlet of the dispenser shall be 14 inches (355 mm) minimum and 19 inches (485 mm) maximum above the finish floor. There shall be a clearance of 1½ inches (38 mm) minimum below the grab bar. Dispensers shall not be of a type that controls delivery or that does not allow continuous paper flow.

604.9.7 Toilet Compartments. Toilet compartments shall comply with 604.8.

605 Urinals

605.1 General. Urinals shall comply with 605.

> **Advisory 605.1 General.** Stall-type urinals provide greater accessibility for a broader range of persons, including people of short stature.

605.2 Height and Depth. Urinals shall be the stall-type or the wall-hung type with the rim 17 inches (430 mm) maximum above the finish floor or ground. Urinals shall be 13½ inches (345 mm) deep minimum measured from the outer face of the urinal rim to the back of the fixture.

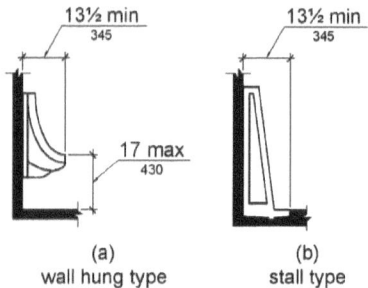

**Figure 605.2
Height and Depth of Urinals**

605.3 Clear Floor Space. A clear floor or ground *space* complying with 305 positioned for forward approach shall be provided.

605.4 Flush Controls. Flush controls shall be hand operated or automatic. Hand operated flush controls shall comply with 309.

606 Lavatories and Sinks

606.1 General. Lavatories and sinks shall comply with 606.

> **Advisory 606.1 General.** If soap and towel dispensers are provided, they must be located within the reach ranges specified in 308. Locate soap and towel dispensers so that they are conveniently usable by a person at the accessible lavatory.

606.2 Clear Floor Space. A clear floor *space* complying with 305, positioned for a forward approach, and knee and toe clearance complying with 306 shall be provided.
EXCEPTIONS: 1. A parallel approach complying with 305 shall be permitted to a kitchen sink in a *space* where a cook top or conventional range is not provided and to wet bars.

CHAPTER 6: PLUMBING ELEMENTS AND FACILITIES

2. A lavatory in a toilet room or bathing *facility* for a single occupant accessed only through a private office and not for *common use* or *public use* shall not be required to provide knee and toe clearance complying with 306.
3. In *residential dwelling units*, cabinetry shall be permitted under lavatories and kitchen sinks provided that all of the following conditions are met:
    (a) the cabinetry can be removed without removal or replacement of the fixture;
    (b) the finish floor extends under the cabinetry; and
    (c) the walls behind and surrounding the cabinetry are finished.
4. A knee clearance of 24 inches (610 mm) minimum above the finish floor or ground shall be permitted at lavatories and sinks used primarily by children 6 through 12 years where the rim or counter surface is 31 inches (785 mm) maximum above the finish floor or ground.
5. A parallel approach complying with 305 shall be permitted to lavatories and sinks used primarily by children 5 years and younger.
6. The dip of the overflow shall not be considered in determining knee and toe clearances.
7. No more than one bowl of a multi-bowl sink shall be required to provide knee and toe clearance complying with 306.

606.3 Height. Lavatories and sinks shall be installed with the front of the higher of the rim or counter surface 34 inches (865 mm) maximum above the finish floor or ground.

EXCEPTIONS: 1. A lavatory in a toilet or bathing *facility* for a single occupant accessed only through a private office and not for *common use* or *public use* shall not be required to comply with 606.3.

2. In *residential dwelling unit* kitchens, sinks that are adjustable to variable heights, 29 inches (735 mm) minimum and 36 inches (915 mm) maximum, shall be permitted where rough-in plumbing permits connections of supply and drain pipes for sinks mounted at the height of 29 inches (735 mm).

606.4 Faucets. Controls for faucets shall comply with 309. Hand-operated metering faucets shall remain open for 10 seconds minimum.

606.5 Exposed Pipes and Surfaces. Water supply and drain pipes under lavatories and sinks shall be insulated or otherwise configured to protect against contact. There shall be no sharp or abrasive surfaces under lavatories and sinks.

607 Bathtubs

607.1 General. Bathtubs shall comply with 607.

607.2 Clearance. Clearance in front of bathtubs shall extend the length of the bathtub and shall be 30 inches (760 mm) wide minimum. A lavatory complying with 606 shall be permitted at the control end of the clearance. Where a permanent seat is provided at the head end of the bathtub, the clearance shall extend 12 inches (305 mm) minimum beyond the wall at the head end of the bathtub.

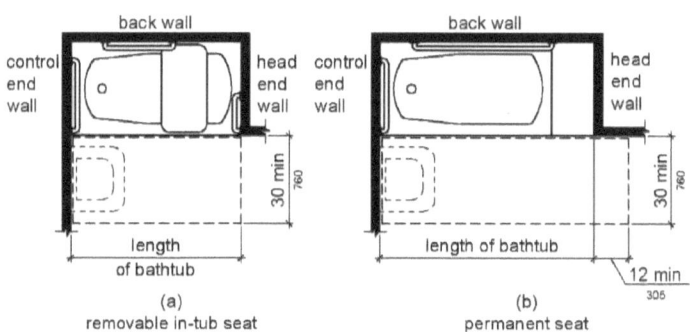

Figure 607.2
Clearance for Bathtubs

607.3 Seat. A permanent seat at the head end of the bathtub or a removable in-tub seat shall be provided. Seats shall comply with 610.

607.4 Grab Bars. Grab bars for bathtubs shall comply with 609 and shall be provided in accordance with 607.4.1 or 607.4.2.

EXCEPTIONS: 1. Grab bars shall not be required to be installed in a bathtub located in a bathing *facility* for a single occupant accessed only through a private office and not for *common use* or *public use* provided that reinforcement has been installed in walls and located so as to permit the installation of grab bars complying with 607.4.

2. In *residential dwelling units*, grab bars shall not be required to be installed in bathtubs located in bathing *facilities* provided that reinforcement has been installed in walls and located so as to permit the installation of grab bars complying with 607.4.

607.4.1 Bathtubs With Permanent Seats. For bathtubs with permanent seats, grab bars shall be provided in accordance with 607.4.1.

607.4.1.1 Back Wall. Two grab bars shall be installed on the back wall, one located in accordance with 609.4 and the other located 8 inches (205 mm) minimum and 10 inches (255 mm) maximum above the rim of the bathtub. Each grab bar shall be installed 15 inches (380 mm) maximum from the head end wall and 12 inches (305 mm) maximum from the control end wall.

607.4.1.2 Control End Wall. A grab bar 24 inches (610 mm) long minimum shall be installed on the control end wall at the front edge of the bathtub.

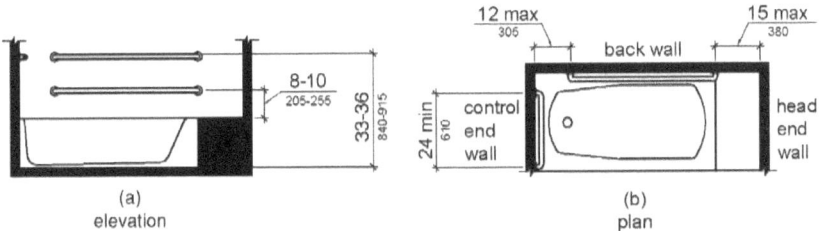

Figure 607.4.1
Grab Bars for Bathtubs with Permanent Seats

607.4.2 Bathtubs Without Permanent Seats. For bathtubs without permanent seats, grab bars shall comply with 607.4.2.

607.4.2.1 Back Wall. Two grab bars shall be installed on the back wall, one located in accordance with 609.4 and other located 8 inches (205 mm) minimum and 10 inches (255 mm) maximum above the rim of the bathtub. Each grab bar shall be 24 inches (610 mm) long minimum and shall be installed 24 inches (610 mm) maximum from the head end wall and 12 inches (305 mm) maximum from the control end wall.

607.4.2.2 Control End Wall. A grab bar 24 inches (610 mm) long minimum shall be installed on the control end wall at the front edge of the bathtub.

607.4.2.3 Head End Wall. A grab bar 12 inches (305 mm) long minimum shall be installed on the head end wall at the front edge of the bathtub.

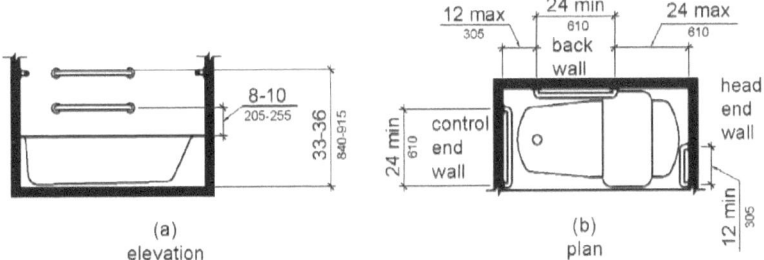

Figure 607.4.2
Grab Bars for Bathtubs with Removable In-Tub Seats

607.5 Controls. Controls, other than drain stoppers, shall be located on an end wall. Controls shall be between the bathtub rim and grab bar, and between the open side of the bathtub and the centerline of the width of the bathtub. Controls shall comply with 309.4.

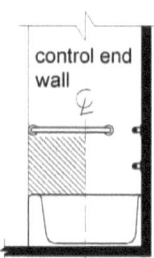

**Figure 607.5
Bathtub Control Location**

607.6 Shower Spray Unit and Water. A shower spray unit with a hose 59 inches (1500 mm) long minimum that can be used both as a fixed-position shower head and as a hand-held shower shall be provided. The shower spray unit shall have an on/off control with a non-positive shut-off. If an adjustable-height shower head on a vertical bar is used, the bar shall be installed so as not to obstruct the use of grab bars. Bathtub shower spray units shall deliver water that is 120°F (49°C) maximum.

> **Advisory 607.6 Shower Spray Unit and Water.** Ensure that hand-held shower spray units are capable of delivering water pressure substantially equivalent to fixed shower heads.

607.7 Bathtub Enclosures. Enclosures for bathtubs shall not obstruct controls, faucets, shower and spray units or obstruct transfer from wheelchairs onto bathtub seats or into bathtubs. Enclosures on bathtubs shall not have tracks installed on the rim of the open face of the bathtub.

608 Shower Compartments

608.1 General. Shower compartments shall comply with 608.

> **Advisory 608.1 General.** Shower stalls that are 60 inches (1525 mm) wide and have no curb may increase the usability of a bathroom because the shower area provides additional maneuvering space.

608.2 Size and Clearances for Shower Compartments. Shower compartments shall have sizes and clearances complying with 608.2.

608.2.1 Transfer Type Shower Compartments. Transfer type shower compartments shall be 36 inches (915 mm) by 36 inches (915 mm) clear inside dimensions measured at the center points of opposing sides and shall have a 36 inch (915 mm) wide minimum entry on the face of the shower compartment. Clearance of 36 inches (915 mm) wide minimum by 48 inches (1220 mm) long minimum measured from the control wall shall be provided.

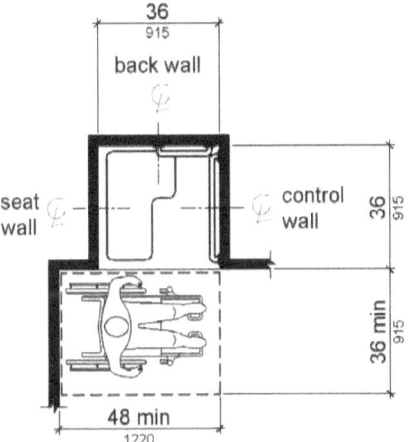

Figure 608.2.1
Transfer Type Shower Compartment Size and Clearance

608.2.2 Standard Roll-In Type Shower Compartments. Standard roll-in type shower compartments shall be 30 inches (760 mm) wide minimum by 60 inches (1525 mm) deep minimum clear inside dimensions measured at center points of opposing sides and shall have a 60 inches (1525 mm) wide minimum entry on the face of the shower compartment.

> **608.2.2.1 Clearance.** A 30 inch (760 mm) wide minimum by 60 inch (1525 mm) long minimum clearance shall be provided adjacent to the open face of the shower compartment.
> **EXCEPTION:** A lavatory complying with 606 shall be permitted on one 30 inch (760 mm) wide minimum side of the clearance provided that it is not on the side of the clearance adjacent to the controls or, where provided, not on the side of the clearance adjacent to the shower seat.

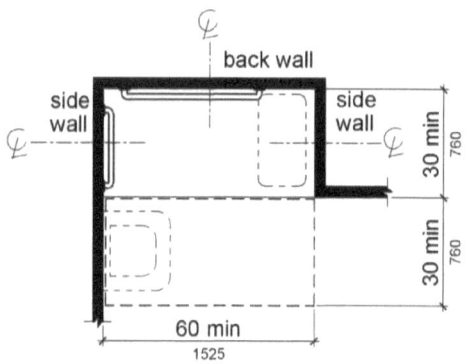

Figure 608.2.2
Standard Roll-In Type Shower Compartment Size and Clearance

608.2.3 Alternate Roll-In Type Shower Compartments. Alternate roll-in type shower compartments shall be 36 inches (915 mm) wide and 60 inches (1525 mm) deep minimum clear inside dimensions measured at center points of opposing sides. A 36 inch (915 mm) wide minimum entry shall be provided at one end of the long side of the compartment.

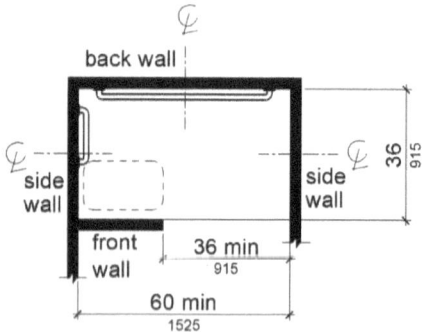

Figure 608.2.3
Alternate Roll-In Type Shower Compartment Size and Clearance

608.3 Grab Bars. Grab bars shall comply with 609 and shall be provided in accordance with 608.3. Where multiple grab bars are used, required horizontal grab bars shall be installed at the same height above the finish floor.
EXCEPTIONS: 1. Grab bars shall not be required to be installed in a shower located in a bathing *facility* for a single occupant accessed only through a private office, and not for *common use* or *public use* provided that reinforcement has been installed in walls and located so as to permit the installation of grab bars complying with 608.3.
2. In *residential dwelling units*, grab bars shall not be required to be installed in showers located in bathing *facilities* provided that reinforcement has been installed in walls and located so as to permit the installation of grab bars complying with 608.3.

608.3.1 Transfer Type Shower Compartments. In transfer type compartments, grab bars shall be provided across the control wall and back wall to a point 18 inches (455 mm) from the control wall.

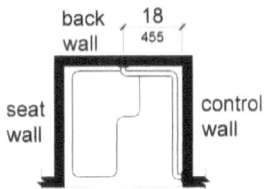

Figure 608.3.1
Grab Bars for Transfer Type Showers

608.3.2 Standard Roll-In Type Shower Compartments. Where a seat is provided in standard roll-in type shower compartments, grab bars shall be provided on the back wall and the side wall opposite the seat. Grab bars shall not be provided above the seat. Where a seat is not provided in standard roll-in type shower compartments, grab bars shall be provided on three walls. Grab bars shall be installed 6 inches (150 mm) maximum from adjacent walls.

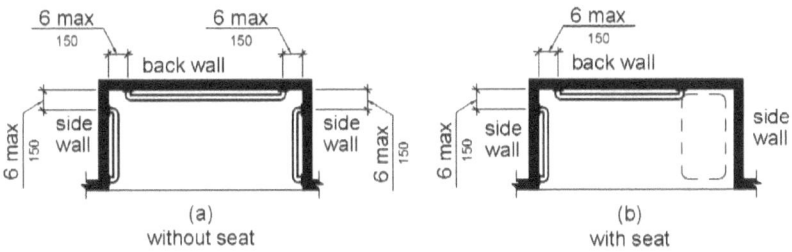

Figure 608.3.2
Grab Bars for Standard Roll-In Type Showers

608.3.3 Alternate Roll-In Type Shower Compartments. In alternate roll-in type shower compartments, grab bars shall be provided on the back wall and the side wall farthest from the compartment entry. Grab bars shall not be provided above the seat. Grab bars shall be installed 6 inches (150 mm) maximum from adjacent walls.

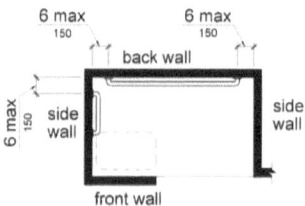

Figure 608.3.3
Grab Bars for Alternate Roll-In Type Showers

608.4 Seats. A folding or non-folding seat shall be provided in transfer type shower compartments. A folding seat shall be provided in roll-in type showers required in *transient lodging* guest rooms with mobility features complying with 806.2. Seats shall comply with 610.

EXCEPTION: In *residential dwelling units*, seats shall not be required in transfer type shower compartments provided that reinforcement has been installed in walls so as to permit the installation of seats complying with 608.4.

608.5 Controls. Controls, faucets, and shower spray units shall comply with 309.4.

608.5.1 Transfer Type Shower Compartments. In transfer type shower compartments, the controls, faucets, and shower spray unit shall be installed on the side wall opposite the seat 38 inches (965 mm) minimum and 48 inches (1220 mm) maximum above the shower floor and shall be located on the control wall 15 inches (380 mm) maximum from the centerline of the seat toward the shower opening.

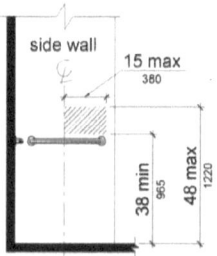

Figure 608.5.1
Transfer Type Shower Compartment Control Location

608.5.2 Standard Roll-In Type Shower Compartments. In standard roll-in type shower compartments, the controls, faucets, and shower spray unit shall be located above the grab bar, but no higher than 48 inches (1220 mm) above the shower floor. Where a seat is provided, the controls, faucets, and shower spray unit shall be installed on the back wall adjacent to the seat wall and shall be located 27 inches (685 mm) maximum from the seat wall.

> **Advisory 608.5.2 Standard Roll-in Type Shower Compartments.** In standard roll-in type showers without seats, the shower head and operable parts can be located on any of the three walls of the shower without adversely affecting accessibility.

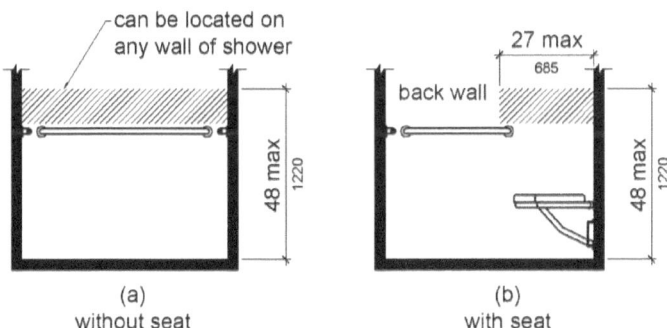

Figure 608.5.2
Standard Roll-In Type Shower Compartment Control Location

608.5.3 Alternate Roll-In Type Shower Compartments. In alternate roll-in type shower compartments, the controls, faucets, and shower spray unit shall be located above the grab bar, but no higher than 48 inches (1220 mm) above the shower floor. Where a seat is provided, the controls, faucets, and shower spray unit shall be located on the side wall adjacent to the seat 27 inches (685 mm) maximum from the side wall behind the seat or shall be located on the back wall opposite the seat 15 inches (380 mm) maximum, left or right, of the centerline of the seat. Where a seat is not provided, the controls, faucets, and shower spray unit shall be installed on the side wall farthest from the compartment entry.

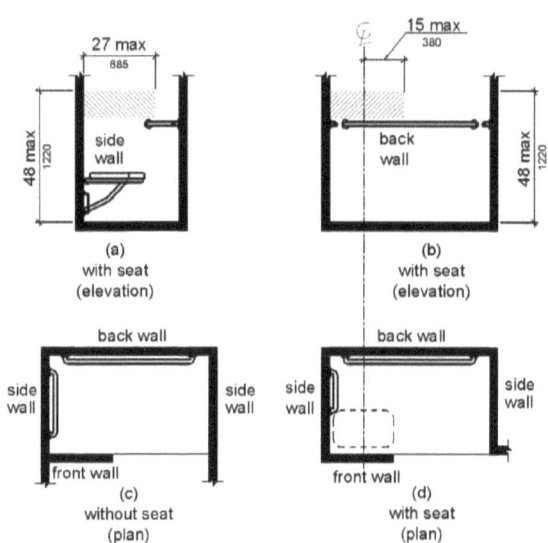

Figure 608.5.3
Alternate Roll-In Type Shower Compartment Control Location

608.6 Shower Spray Unit and Water. A shower spray unit with a hose 59 inches (1500 mm) long minimum that can be used both as a fixed-position shower head and as a hand-held shower shall be provided. The shower spray unit shall have an on/off control with a non-positive shut-off. If an adjustable-height shower head on a vertical bar is used, the bar shall be installed so as not to obstruct the use of grab bars. Shower spray units shall deliver water that is 120°F (49°C) maximum.
    EXCEPTION: A fixed shower head located at 48 inches (1220 mm) maximum above the shower finish floor shall be permitted instead of a hand-held spray unit in *facilities* that are not medical care *facilities*, long-term care *facilities*, *transient lodging* guest rooms, or *residential dwelling units*.

> **Advisory 608.6 Shower Spray Unit and Water.** Ensure that hand-held shower spray units are capable of delivering water pressure substantially equivalent to fixed shower heads.

608.7 Thresholds. Thresholds in roll-in type shower compartments shall be ½ inch (13 mm) high maximum in accordance with 303. In transfer type shower compartments, thresholds ½ inch (13 mm) high maximum shall be beveled, rounded, or vertical.
    EXCEPTION: A threshold 2 inches (51 mm) high maximum shall be permitted in transfer type shower compartments in existing *facilities* where provision of a ½ inch (13 mm) high threshold would disturb the structural reinforcement of the floor slab.

608.8 Shower Enclosures. Enclosures for shower compartments shall not obstruct controls, faucets, and shower spray units or obstruct transfer from wheelchairs onto shower seats.

609 Grab Bars

609.1 General. Grab bars in toilet *facilities* and bathing *facilities* shall comply with 609.

609.2 Cross Section. Grab bars shall have a cross section complying with 609.2.1 or 609.2.2.

    609.2.1 Circular Cross Section. Grab bars with circular cross sections shall have an outside diameter of 1¼ inches (32 mm) minimum and 2 inches (51 mm) maximum.

    609.2.2 Non-Circular Cross Section. Grab bars with non-circular cross sections shall have a cross-section dimension of 2 inches (51 mm) maximum and a perimeter dimension of 4 inches (100 mm) minimum and 4.8 inches (120 mm) maximum.

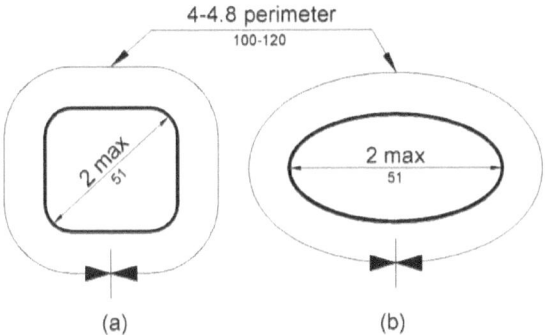

Figure 609.2.2
Grab Bar Non-Circular Cross Section

609.3 Spacing. The *space* between the wall and the grab bar shall be 1½ inches (38 mm). The *space* between the grab bar and projecting objects below and at the ends shall be 1½ inches (38 mm) minimum. The *space* between the grab bar and projecting objects above shall be 12 inches (305 mm) minimum.

    EXCEPTION: The *space* between the grab bars and shower controls, shower fittings, and other grab bars above shall be permitted to be 1½ inches (38 mm) minimum.

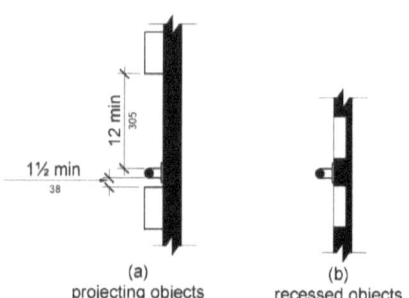

**Figure 609.3
Spacing of Grab Bars**

609.4 Position of Grab Bars. Grab bars shall be installed in a horizontal position, 33 inches (840 mm) minimum and 36 inches (915 mm) maximum above the finish floor measured to the top of the gripping surface, except that at water closets for *children's use* complying with 604.9, grab bars shall be installed in a horizontal position 18 inches (455 mm) minimum and 27 inches (685 mm) maximum above the finish floor measured to the top of the gripping surface. The height of the lower grab bar on the back wall of a bathtub shall comply with 607.4.1.1 or 607.4.2.1.

609.5 Surface Hazards. Grab bars and any wall or other surfaces adjacent to grab bars shall be free of sharp or abrasive *elements* and shall have rounded edges.

609.6 Fittings. Grab bars shall not rotate within their fittings.

609.7 Installation. Grab bars shall be installed in any manner that provides a gripping surface at the specified locations and that does not obstruct the required clear floor *space*.

609.8 Structural Strength. Allowable stresses shall not be exceeded for materials used when a vertical or horizontal force of 250 pounds (1112 N) is applied at any point on the grab bar, fastener, mounting device, or supporting structure.

610 Seats

610.1 General. Seats in bathtubs and shower compartments shall comply with 610.

610.2 Bathtub Seats. The top of bathtub seats shall be 17 inches (430 mm) minimum and 19 inches (485 mm) maximum above the bathroom finish floor. The depth of a removable in-tub seat shall be 15 inches (380 mm) minimum and 16 inches (405 mm) maximum. The seat shall be capable of secure placement. Permanent seats at the head end of the bathtub shall be 15 inches (380 mm) deep minimum and shall extend from the back wall to or beyond the outer edge of the bathtub.

CHAPTER 6: PLUMBING ELEMENTS AND FACILITIES — TECHNICAL

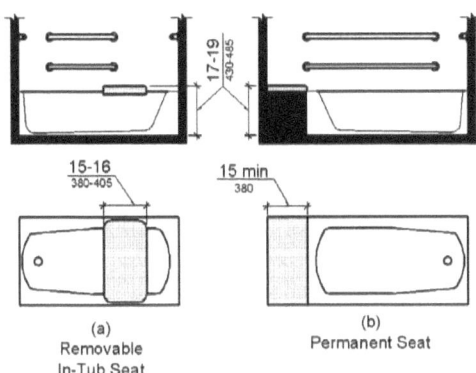

Figure 610.2
Bathtub Seats

610.3 Shower Compartment Seats. Where a seat is provided in a standard roll-in shower compartment, it shall be a folding type, shall be installed on the side wall adjacent to the controls, and shall extend from the back wall to a point within 3 inches (75 mm) of the compartment entry. Where a seat is provided in an alternate roll-in type shower compartment, it shall be a folding type, shall be installed on the front wall opposite the back wall, and shall extend from the adjacent side wall to a point within 3 inches (75 mm) of the compartment entry. In transfer-type showers, the seat shall extend from the back wall to a point within 3 inches (75 mm) of the compartment entry. The top of the seat shall be 17 inches (430 mm) minimum and 19 inches (485 mm) maximum above the bathroom finish floor. Seats shall comply with 610.3.1 or 610.3.2.

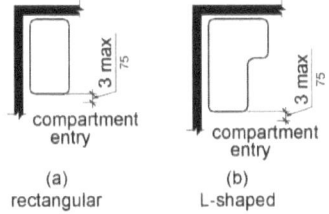

Figure 610.3
Extent of Seat

610.3.1 Rectangular Seats. The rear edge of a rectangular seat shall be 2½ inches (64 mm) maximum and the front edge 15 inches (380 mm) minimum and 16 inches (405 mm) maximum from the seat wall. The side edge of the seat shall be 1½ inches (38 mm) maximum from the adjacent wall.

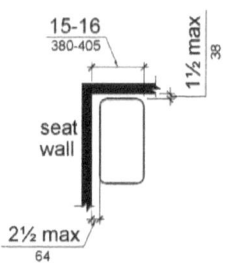

Figure 610.3.1
Rectangular Shower Seat

610.3.2 L-Shaped Seats. The rear edge of an L-shaped seat shall be 2½ inches (64 mm) maximum and the front edge 15 inches (380 mm) minimum and 16 inches (405 mm) maximum from the seat wall. The rear edge of the "L" portion of the seat shall be 1½ inches (38 mm) maximum from the wall and the front edge shall be 14 inches (355 mm) minimum and 15 inches (380 mm) maximum from the wall. The end of the "L" shall be 22 inches (560 mm) minimum and 23 inches maximum (585 mm) from the main seat wall.

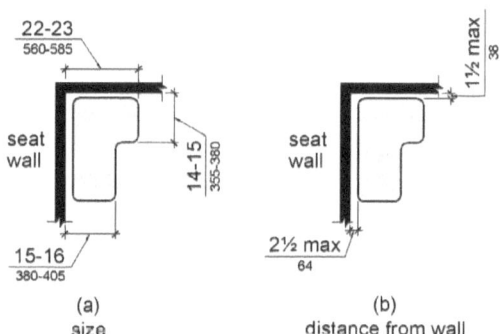

(a)
size

(b)
distance from wall

Figure 610.3.2
L-Shaped Shower Seat

610.4 Structural Strength. Allowable stresses shall not be exceeded for materials used when a vertical or horizontal force of 250 pounds (1112 N) is applied at any point on the seat, fastener, mounting device, or supporting structure.

611 Washing Machines and Clothes Dryers

611.1 General. Washing machines and clothes dryers shall comply with 611.

611.2 Clear Floor Space. A clear floor or ground *space* complying with 305 positioned for parallel approach shall be provided. The clear floor or ground *space* shall be centered on the appliance.

611.3 Operable Parts. *Operable parts*, including doors, lint screens, and detergent and bleach compartments shall comply with 309.

611.4 Height. Top loading machines shall have the door to the laundry compartment located 36 inches (915 mm) maximum above the finish floor. Front loading machines shall have the bottom of the opening to the laundry compartment located 15 inches (380 mm) minimum and 36 inches (915 mm) maximum above the finish floor.

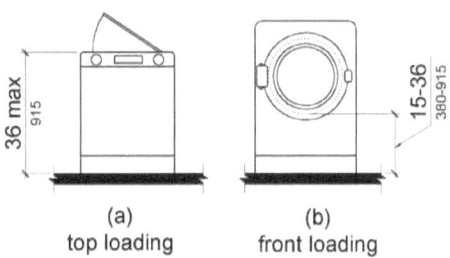

(a) top loading

(b) front loading

Figure 611.4
Height of Laundry Compartment Opening

612 Saunas and Steam Rooms

612.1 General. Saunas and steam rooms shall comply with 612.

612.2 Bench. Where seating is provided in saunas and steam rooms, at least one bench shall comply with 903. Doors shall not swing into the clear floor *space* required by 903.2.

    EXCEPTION: A readily removable bench shall be permitted to obstruct the turning *space* required by 612.3 and the clear floor or ground *space* required by 903.2.

612.3 Turning Space. A turning *space* complying with 304 shall be provided within saunas and steam rooms.

CHAPTER 7: COMMUNICATION ELEMENTS AND FEATURES

701 General

701.1 Scope. The provisions of Chapter 7 shall apply where required by Chapter 2 or where referenced by a requirement in this document.

702 Fire Alarm Systems

702.1 General. Fire alarm systems shall have permanently installed audible and visible alarms complying with NFPA 72 (1999 or 2002 edition) (incorporated by reference, see "Referenced Standards" in Chapter 1), except that the maximum allowable sound level of audible notification appliances complying with section 4-3.2.1 of NFPA 72 (1999 edition) shall have a sound level no more than 110 dB at the minimum hearing distance from the audible appliance. In addition, alarms in guest rooms required to provide communication features shall comply with sections 4-3 and 4-4 of NFPA 72 (1999 edition) or sections 7.4 and 7.5 of NFPA 72 (2002 edition).

> **EXCEPTION:** Fire alarm systems in medical care *facilities* shall be permitted to be provided in accordance with industry practice.

703 Signs

703.1 General. Signs shall comply with 703. Where both visual and *tactile characters* are required, either one sign with both visual and *tactile characters*, or two separate signs, one with visual, and one with *tactile characters*, shall be provided.

703.2 Raised Characters. Raised *characters* shall comply with 703.2 and shall be duplicated in braille complying with 703.3. Raised *characters* shall be installed in accordance with 703.4.

> **Advisory 703.2 Raised Characters.** Signs that are designed to be read by touch should not have sharp or abrasive edges.

703.2.1 Depth. Raised *characters* shall be 1/32 inch (0.8 mm) minimum above their background.

703.2.2 Case. *Characters* shall be uppercase.

703.2.3 Style. *Characters* shall be sans serif. *Characters* shall not be italic, oblique, script, highly decorative, or of other unusual forms.

703.2.4 Character Proportions. *Characters* shall be selected from fonts where the width of the uppercase letter "O" is 55 percent minimum and 110 percent maximum of the height of the uppercase letter "I".

703.2.5 Character Height. *Character* height measured vertically from the baseline of the *character* shall be 5/8 inch (16 mm) minimum and 2 inches (51 mm) maximum based on the height of the uppercase letter "I".

EXCEPTION: Where separate raised and visual *characters* with the same information are provided, raised *character* height shall be permitted to be ½ inch (13 mm) minimum.

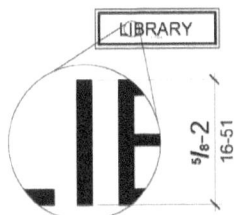

Figure 703.2.5
Height of Raised Characters

703.2.6 Stroke Thickness. Stroke thickness of the uppercase letter "I" shall be 15 percent maximum of the height of the *character*.

703.2.7 Character Spacing. *Character* spacing shall be measured between the two closest points of adjacent raised *characters* within a message, excluding word *spaces*. Where *characters* have rectangular cross sections, spacing between individual raised *characters* shall be 1/8 inch (3.2 mm) minimum and 4 times the raised *character* stroke width maximum. Where *characters* have other cross sections, spacing between individual raised *characters* shall be 1/16 inch (1.6 mm) minimum and 4 times the raised *character* stroke width maximum at the base of the cross sections, and 1/8 inch (3.2 mm) minimum and 4 times the raised *character* stroke width maximum at the top of the cross sections. *Characters* shall be separated from raised borders and decorative *elements* 3/8 inch (9.5 mm) minimum.

703.2.8 Line Spacing. Spacing between the baselines of separate lines of raised *characters* within a message shall be 135 percent minimum and 170 percent maximum of the raised *character* height.

703.3 Braille. Braille shall be contracted (Grade 2) and shall comply with 703.3 and 703.4.

703.3.1 Dimensions and Capitalization. Braille dots shall have a domed or rounded shape and shall comply with Table 703.3.1. The indication of an uppercase letter or letters shall only be used before the first word of sentences, proper nouns and names, individual letters of the alphabet, initials, and acronyms.

TECHNICAL   CHAPTER 7: COMMUNICATION ELEMENTS AND FEATURES

Table 703.3.1 Braille Dimensions

Measurement Range	Minimum in Inches Maximum in Inches
Dot base diameter	0.059 (1.5 mm) to 0.063 (1.6 mm)
Distance between two dots in the same cell[1]	0.090 (2.3 mm) to 0.100 (2.5 mm)
Distance between corresponding dots in adjacent cells[1]	0.241 (6.1 mm) to 0.300 (7.6 mm)
Dot height	0.025 (0.6 mm) to 0.037 (0.9 mm)
Distance between corresponding dots from one cell directly below[1]	0.395 (10 mm) to 0.400 (10.2 mm)

1. Measured center to center.

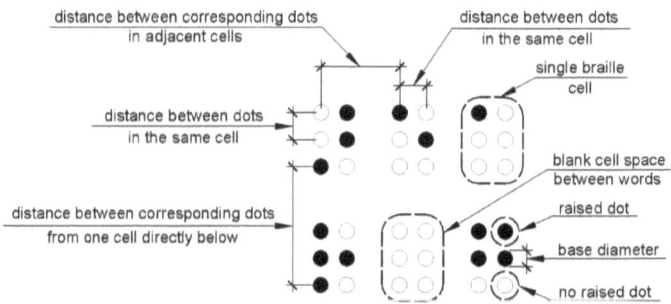

Figure 703.3.1
Braille Measurement

703.3.2 Position. Braille shall be positioned below the corresponding text. If text is multi-lined, braille shall be placed below the entire text. Braille shall be separated 3/8 inch (9.5 mm) minimum from any other *tactile characters* and 3/8 inch (9.5 mm) minimum from raised borders and decorative *elements*.

EXCEPTION: Braille provided on elevator car controls shall be separated 3/16 inch (4.8 mm) minimum and shall be located either directly below or adjacent to the corresponding raised *characters* or symbols.

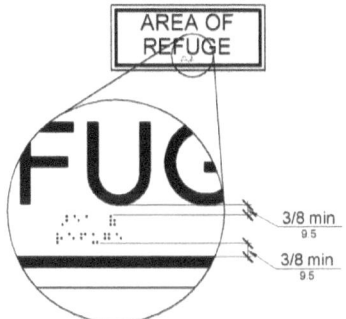

Figure 703.3.2
Position of Braille

703.4 Installation Height and Location. Signs with *tactile characters* shall comply with 703.4.

703.4.1 Height Above Finish Floor or Ground. *Tactile characters* on signs shall be located 48 inches (1220 mm) minimum above the finish floor or ground surface, measured from the baseline of the lowest *tactile character* and 60 inches (1525 mm) maximum above the finish floor or ground surface, measured from the baseline of the highest *tactile character*.

EXCEPTION: *Tactile characters* for elevator car controls shall not be required to comply with 703.4.1.

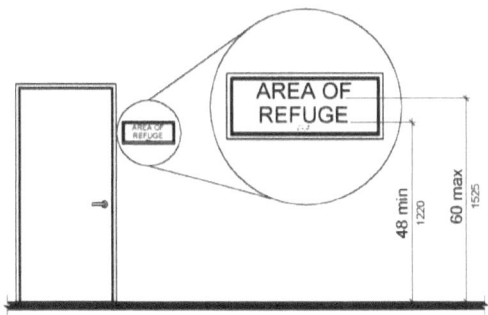

Figure 703.4.1
Height of Tactile Characters Above Finish Floor or Ground

703.4.2 Location. Where a *tactile* sign is provided at a door, the sign shall be located alongside the door at the latch side. Where a *tactile* sign is provided at double doors with one active leaf, the sign shall be located on the inactive leaf. Where a *tactile* sign is provided at double doors with two active leafs, the sign shall be located to the right of the right hand door. Where there is no wall *space* at the latch side of a single door or at the right side of double doors, signs shall be located on the nearest adjacent wall. Signs containing *tactile characters* shall be located so that a clear floor *space* of 18 inches (455 mm) minimum by 18 inches (455 mm) minimum, centered on the *tactile characters*, is provided beyond the arc of any door swing between the closed position and 45 degree open position.

EXCEPTION: Signs with *tactile characters* shall be permitted on the push side of doors with closers and without hold-open devices.

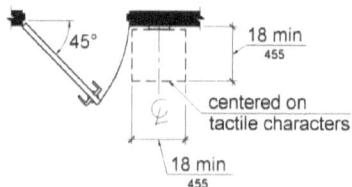

Figure 703.4.2
Location of Tactile Signs at Doors

703.5 Visual Characters. Visual *characters* shall comply with 703.5.

EXCEPTION: Where visual *characters* comply with 703.2 and are accompanied by braille complying with 703.3, they shall not be required to comply with 703.5.2 through 703.5.9.

703.5.1 Finish and Contrast. *Characters* and their background shall have a non-glare finish. *Characters* shall contrast with their background with either light *characters* on a dark background or dark *characters* on a light background.

> **Advisory 703.5.1 Finish and Contrast.** Signs are more legible for persons with low vision when characters contrast as much as possible with their background. Additional factors affecting the ease with which the text can be distinguished from its background include shadows cast by lighting sources, surface glare, and the uniformity of the text and its background colors and textures.

703.5.2 Case. *Characters* shall be uppercase or lowercase or a combination of both.

703.5.3 Style. *Characters* shall be conventional in form. *Characters* shall not be italic, oblique, script, highly decorative, or of other unusual forms.

703.5.4 Character Proportions. *Characters* shall be selected from fonts where the width of the uppercase letter "O" is 55 percent minimum and 110 percent maximum of the height of the uppercase letter "I".

703.5.5 Character Height. Minimum *character* height shall comply with Table 703.5.5. Viewing distance shall be measured as the horizontal distance between the *character* and an obstruction preventing further approach towards the sign. *Character* height shall be based on the uppercase letter "I".

Table 703.5.5 Visual Character Height

Height to Finish Floor or Ground From Baseline of Character	Horizontal Viewing Distance	Minimum Character Height
40 inches (1015 mm) to less than or equal to 70 inches (1780 mm)	less than 72 inches (1830 mm)	5/8 inch (16 mm)
	72 inches (1830 mm) and greater	5/8 inch (16 mm), plus 1/8 inch (3.2 mm) per foot (305 mm) of viewing distance above 72 inches (1830 mm)
Greater than 70 inches (1780 mm) to less than or equal to 120 inches (3050 mm)	less than 180 inches (4570 mm)	2 inches (51 mm)
	180 inches (4570 mm) and greater	2 inches (51 mm), plus 1/8 inch (3.2 mm) per foot (305 mm) of viewing distance above 180 inches (4570 mm)
greater than 120 inches (3050 mm)	less than 21 feet (6400 mm)	3 inches (75 mm)
	21 feet (6400 mm) and greater	3 inches (75 mm), plus 1/8 inch (3.2 mm) per foot (305 mm) of viewing distance above 21 feet (6400 mm)

703.5.6 Height From Finish Floor or Ground. Visual *characters* shall be 40 inches (1015 mm) minimum above the finish floor or ground.

EXCEPTION: Visual *characters* indicating elevator car controls shall not be required to comply with 703.5.6.

703.5.7 Stroke Thickness. Stroke thickness of the uppercase letter "I" shall be 10 percent minimum and 30 percent maximum of the height of the *character*.

703.5.8 Character Spacing. *Character* spacing shall be measured between the two closest points of adjacent *characters*, excluding word *spaces*. Spacing between individual *characters* shall be 10 percent minimum and 35 percent maximum of *character* height.

703.5.9 Line Spacing. Spacing between the baselines of separate lines of *characters* within a message shall be 135 percent minimum and 170 percent maximum of the *character* height.

703.6 Pictograms. *Pictograms* shall comply with 703.6.

703.6.1 Pictogram Field. *Pictograms* shall have a field height of 6 inches (150 mm) minimum. *Characters* and braille shall not be located in the *pictogram* field.

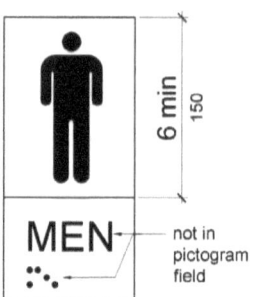

Figure 703.6.1
Pictogram Field

703.6.2 Finish and Contrast. *Pictograms* and their field shall have a non-glare finish. *Pictograms* shall contrast with their field with either a light *pictogram* on a dark field or a dark *pictogram* on a light field.

> **Advisory 703.6.2 Finish and Contrast.** Signs are more legible for persons with low vision when characters contrast as much as possible with their background. Additional factors affecting the ease with which the text can be distinguished from its background include shadows cast by lighting sources, surface glare, and the uniformity of the text and background colors and textures.

703.6.3 Text Descriptors. *Pictograms* shall have text descriptors located directly below the *pictogram* field. Text descriptors shall comply with 703.2, 703.3 and 703.4.

703.7 Symbols of Accessibility. Symbols of *accessibility* shall comply with 703.7.

703.7.1 Finish and Contrast. Symbols of *accessibility* and their background shall have a non-glare finish. Symbols of *accessibility* shall contrast with their background with either a light symbol on a dark background or a dark symbol on a light background.

> **Advisory 703.7.1 Finish and Contrast.** Signs are more legible for persons with low vision when characters contrast as much as possible with their background. Additional factors affecting the ease with which the text can be distinguished from its background include shadows cast by lighting sources, surface glare, and the uniformity of the text and background colors and textures.

CHAPTER 7: COMMUNICATION ELEMENTS AND FEATURES   TECHNICAL

703.7.2 Symbols.

703.7.2.1 International Symbol of Accessibility. The International Symbol of *Accessibility* shall comply with Figure 703.7.2.1.

Figure 703.7.2.1
International Symbol of Accessibility

703.7.2.2 International Symbol of TTY. The International Symbol of *TTY* shall comply with Figure 703.7.2.2.

Figure 703.7.2.2
International Symbol of TTY

703.7.2.3 Volume Control Telephones. Telephones with a volume control shall be identified by a *pictogram* of a telephone handset with radiating sound waves on a square field such as shown in Figure 703.7.2.3.

Figure 703.7.2.3
Volume Control Telephone

703.7.2.4 Assistive Listening Systems. *Assistive listening systems* shall be identified by the International Symbol of Access for Hearing Loss complying with Figure 703.7.2.4.

Figure 703.7.2.4
International Symbol of Access for Hearing Loss

704 Telephones

704.1 General. Public telephones shall comply with 704.

704.2 Wheelchair Accessible Telephones. Wheelchair *accessible* telephones shall comply with 704.2.

704.2.1 Clear Floor or Ground Space. A clear floor or ground *space* complying with 305 shall be provided. The clear floor or ground *space* shall not be obstructed by bases, enclosures, or seats.

> **Advisory 704.2.1 Clear Floor or Ground Space.** Because clear floor and ground space is required to be unobstructed, telephones, enclosures and related telephone book storage cannot encroach on the required clear floor or ground space and must comply with the provisions for protruding objects. (See Section 307.)

704.2.1.1 Parallel Approach. Where a parallel approach is provided, the distance from the edge of the telephone enclosure to the face of the telephone unit shall be 10 inches (255 mm) maximum.

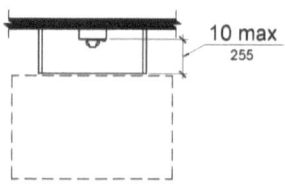

Figure 704.2.1.1
Parallel Approach to Telephone

704.2.1.2 Forward Approach. Where a forward approach is provided, the distance from the front edge of a counter within the telephone enclosure to the face of the telephone unit shall be 20 inches (510 mm) maximum.

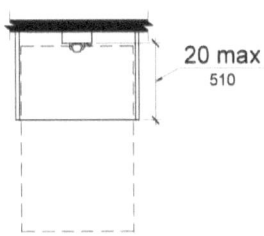

Figure 704.2.1.2
Forward Approach to Telephone

704.2.2 Operable Parts. *Operable parts* shall comply with 309. Telephones shall have push-button controls where such service is available.

704.2.3 Telephone Directories. Telephone directories, where provided, shall be located in accordance with 309.

704.2.4 Cord Length. The cord from the telephone to the handset shall be 29 inches (735 mm) long minimum.

704.3 Volume Control Telephones. Public telephones required to have volume controls shall be equipped with a receive volume control that provides a gain adjustable up to 20 dB minimum. For incremental volume control, provide at least one intermediate step of 12 dB of gain minimum. An automatic reset shall be provided.

> **Advisory 704.3 Volume Control Telephones.** Amplifiers on pay phones are located in the base or the handset or are built into the telephone. Most are operated by pressing a button or key. If the microphone in the handset is not being used, a mute button that temporarily turns off the microphone can also reduce the amount of background noise which the person hears in the earpiece. If a volume adjustment is provided that allows the user to set the level anywhere from the base volume to the upper requirement of 20 dB, there is no need to specify a lower limit. If a stepped volume control is provided, one of the intermediate levels must provide 12 dB of gain. Consider compatibility issues when matching an amplified handset with a phone or phone system. Amplified handsets that can be switched with pay telephone handsets are available. Portable and in-line amplifiers can be used with some phones but are not practical at most public phones covered by these requirements.

704.4 TTYs. *TTYs* required at a public pay telephone shall be permanently affixed within, or adjacent to, the telephone enclosure. Where an acoustic coupler is used, the telephone cord shall be sufficiently long to allow connection of the *TTY* and the telephone receiver.

> **Advisory 704.4 TTYs.** Ensure that sufficient electrical service is available where TTYs are to be installed.

704.4.1 Height. When in use, the touch surface of *TTY* keypads shall be 34 inches (865 mm) minimum above the finish floor.

    EXCEPTION: Where seats are provided, *TTYs* shall not be required to comply with 704.4.1.

> **Advisory 704.4.1 Height.** A telephone with a TTY installed underneath cannot also be a wheelchair accessible telephone because the required 34 inches (865 mm) minimum keypad height can causes the highest operable part of the telephone, usually the coin slot, to exceed the maximum permitted side and forward reach ranges. (See Section 308).
>
> **Advisory 704.4.1 Height Exception.** While seats are not required at TTYs, reading and typing at a TTY is more suited to sitting than standing. Facilities that often provide seats at TTY's include, but are not limited to, airports and other passenger terminals or stations, courts, art galleries, and convention centers.

704.5 TTY Shelf. Public pay telephones required to accommodate portable *TTYs* shall be equipped with a shelf and an electrical outlet within or adjacent to the telephone enclosure. The telephone handset shall be capable of being placed flush on the surface of the shelf. The shelf shall be capable of accommodating a *TTY* and shall have 6 inches (150 mm) minimum vertical clearance above the area where the *TTY* is to be placed.

705 Detectable Warnings

705.1 General. *Detectable warnings* shall consist of a surface of truncated domes and shall comply with 705.

    705.1.1 Dome Size. Truncated domes in a *detectable warning* surface shall have a base diameter of 0.9 inch (23 mm) minimum and 1.4 inches (36 mm) maximum, a top diameter of 50 percent of the base diameter minimum to 65 percent of the base diameter maximum, and a height of 0.2 inch (5.1 mm).

    705.1.2 Dome Spacing. Truncated domes in a *detectable warning* surface shall have a center-to-center spacing of 1.6 inches (41 mm) minimum and 2.4 inches (61 mm) maximum, and a base-to-base spacing of 0.65 inch (17 mm) minimum, measured between the most adjacent domes on a square grid.

    705.1.3 Contrast. *Detectable warning* surfaces shall contrast visually with adjacent walking surfaces either light-on-dark, or dark-on-light.

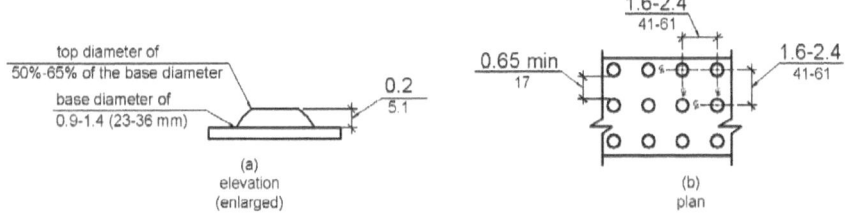

Figure 705.1
Size and Spacing of Truncated Domes

705.2 Platform Edges. Detectable warning surfaces at platform boarding edges shall be 24 inches (610 mm) wide and shall extend the full length of the *public use* areas of the platform.

706 Assistive Listening Systems

706.1 General. *Assistive listening systems* required in *assembly areas* shall comply with 706.

> **Advisory 706.1 General.** Assistive listening systems are generally categorized by their mode of transmission. There are hard-wired systems and three types of wireless systems: induction loop, infrared, and FM radio transmission. Each has different advantages and disadvantages that can help determine which system is best for a given application. For example, an FM system may be better than an infrared system in some open-air assemblies since infrared signals are less effective in sunlight. On the other hand, an infrared system is typically a better choice than an FM system where confidential transmission is important because it will be contained within a given space.
>
> The technical standards for assistive listening systems describe minimum performance levels for volume, interference, and distortion. Sound pressure levels (SPL), expressed in decibels, measure output sound volume. Signal-to-noise ratio (SNR or S/N), also expressed in decibels, represents the relationship between the loudness of a desired sound (the signal) and the background noise in a space or piece of equipment. The higher the SNR, the more intelligible the signal. The peak clipping level limits the distortion in signal output produced when high-volume sound waves are manipulated to serve assistive listening devices.
>
> Selecting or specifying an effective assistive listening system for a large or complex venue requires assistance from a professional sound engineer. The Access Board has published technical assistance on assistive listening devices and systems.

706.2 Receiver Jacks. Receivers required for use with an *assistive listening system* shall include a 1/8 inch (3.2 mm) standard mono jack.

706.3 Receiver Hearing-Aid Compatibility. Receivers required to be hearing-aid compatible shall interface with telecoils in hearing aids through the provision of neckloops.

> **Advisory 706.3 Receiver Hearing-Aid Compatibility.** Neckloops and headsets that can be worn as neckloops are compatible with hearing aids. Receivers that are not compatible include earbuds, which may require removal of hearing aids, earphones, and headsets that must be worn over the ear, which can create disruptive interference in the transmission and can be uncomfortable for people wearing hearing aids.

706.4 Sound Pressure Level. *Assistive listening systems* shall be capable of providing a sound pressure level of 110 dB minimum and 118 dB maximum with a dynamic range on the volume control of 50 dB.

706.5 Signal-to-Noise Ratio. The signal-to-noise ratio for internally generated noise in *assistive listening systems* shall be 18 dB minimum.

706.6 Peak Clipping Level. Peak clipping shall not exceed 18 dB of clipping relative to the peaks of speech.

707 Automatic Teller Machines and Fare Machines

> **Advisory 707 Automatic Teller Machines and Fare Machines.** Interactive transaction machines (ITMs), other than ATMs, are not covered by Section 707. However, for entities covered by the ADA, the Department of Justice regulations that implement the ADA provide additional guidance regarding the relationship between these requirements and elements that are not directly addressed by these requirements. Federal procurement law requires that ITMs purchased by the Federal government comply with standards issued by the Access Board under Section 508 of the Rehabilitation Act of 1973, as amended. This law covers a variety of products, including computer hardware and software, websites, phone systems, fax machines, copiers, and similar technologies. For more information on Section 508 consult the Access Board's website at www.access-board.gov.

707.1 General. Automatic teller machines and fare machines shall comply with 707.

> **Advisory 707.1 General.** If farecards have one tactually distinctive corner they can be inserted with greater accuracy. Token collection devices that are designed to accommodate tokens which are perforated can allow a person to distinguish more readily between tokens and common coins. Place accessible gates and fare vending machines in close proximity to other accessible elements when feasible so the facility is easier to use.

707.2 Clear Floor or Ground Space. A clear floor or ground *space* complying with 305 shall be provided.

    EXCEPTION: Clear floor or ground *space* shall not be required at drive-up only automatic teller machines and fare machines.

707.3 Operable Parts. *Operable parts* shall comply with 309. Unless a clear or correct key is provided, each *operable part* shall be able to be differentiated by sound or touch, without activation.
  EXCEPTION: Drive-up only automatic teller machines and fare machines shall not be required to comply with 309.2 and 309.3.

707.4 Privacy. Automatic teller machines shall provide the opportunity for the same degree of privacy of input and output available to all individuals.

> **Advisory 707.4 Privacy.** In addition to people who are blind or visually impaired, people with limited reach who use wheelchairs or have short stature, who cannot effectively block the ATM screen with their bodies, may prefer to use speech output. Speech output users can benefit from an option to render the visible screen blank, thereby affording them greater personal security and privacy.

707.5 Speech Output. Machines shall be speech enabled. Operating instructions and orientation, visible transaction prompts, user input verification, error messages, and all displayed information for full use shall be *accessible* to and independently usable by individuals with vision impairments. Speech shall be delivered through a mechanism that is readily available to all users, including but not limited to, an industry standard connector or a telephone handset. Speech shall be recorded or digitized human, or synthesized.
  EXCEPTIONS: 1. Audible tones shall be permitted instead of speech for visible output that is not displayed for security purposes, including but not limited to, asterisks representing personal identification numbers.
  2. Advertisements and other similar information shall not be required to be audible unless they convey information that can be used in the transaction being conducted.
  3. Where speech synthesis cannot be supported, dynamic alphabetic output shall not be required to be audible.

> **Advisory 707.5 Speech Output.** If an ATM provides additional functions such as dispensing coupons, selling theater tickets, or providing copies of monthly statements, all such functions must be available to customers using speech output. To avoid confusion at the ATM, the method of initiating the speech mode should be easily discoverable and should not require specialized training. For example, if a telephone handset is provided, lifting the handset can initiate the speech mode.

707.5.1 User Control. Speech shall be capable of being repeated or interrupted. Volume control shall be provided for the speech function.
  EXCEPTION: Speech output for any single function shall be permitted to be automatically interrupted when a transaction is selected.

707.5.2 Receipts. Where receipts are provided, speech output devices shall provide audible balance inquiry information, error messages, and all other information on the printed receipt necessary to complete or verify the transaction.
  EXCEPTIONS: 1. Machine location, date and time of transaction, customer account number, and the machine identifier shall not be required to be audible.

| TECHNICAL | CHAPTER 7: COMMUNICATION ELEMENTS AND FEATURES |

   2. Information on printed receipts that duplicates information available on-screen shall not be required to be presented in the form of an audible receipt.
   3. Printed copies of bank statements and checks shall not be required to be audible.

707.6 Input. Input devices shall comply with 707.6.

707.6.1 Input Controls. At least one *tactilely* discernible input control shall be provided for each function. Where provided, key surfaces not on active areas of display screens, shall be raised above surrounding surfaces. Where membrane keys are the only method of input, each shall be *tactilely* discernable from surrounding surfaces and adjacent keys.

707.6.2 Numeric Keys. Numeric keys shall be arranged in a 12-key ascending or descending telephone keypad layout. The number five key shall be *tactilely* distinct from the other keys.

> **Advisory 707.6.2 Numeric Keys.** Telephone keypads and computer keyboards differ in one significant feature, ascending versus descending numerical order. Both types of keypads are acceptable, provided the computer-style keypad is organized similarly to the number pad located at the right on most computer keyboards, and does not resemble the line of numbers located above the computer keys.

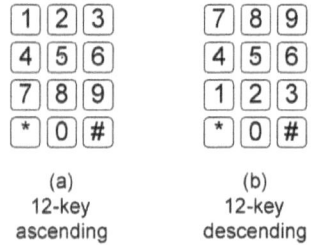

(a) 12-key ascending   (b) 12-key descending

Figure 707.6.2
Numeric Key Layout

707.6.3 Function Keys. Function keys shall comply with 707.6.3.

707.6.3.1 Contrast. Function keys shall contrast visually from background surfaces. *Characters* and symbols on key surfaces shall contrast visually from key surfaces. Visual contrast shall be either light-on-dark or dark-on-light.
   EXCEPTION: *Tactile* symbols required by 707.6.3.2 shall not be required to comply with 707.6.3.1.

707.6.3.2 Tactile Symbols. Function key surfaces shall have *tactile* symbols as follows: Enter or Proceed key: raised circle; Clear or Correct key: raised left arrow; Cancel key: raised letter ex; Add Value key: raised plus sign; Decrease Value key: raised minus sign.

707.7 Display Screen. The display screen shall comply with 707.7.
EXCEPTION: Drive-up only automatic teller machines and fare machines shall not be required to comply with 707.7.1.

707.7.1 Visibility. The display screen shall be visible from a point located 40 inches (1015 mm) above the center of the clear floor *space* in front of the machine.

707.7.2 Characters. *Characters* displayed on the screen shall be in a sans serif font. *Characters* shall be 3/16 inch (4.8 mm) high minimum based on the uppercase letter "I". *Characters* shall contrast with their background with either light *characters* on a dark background or dark *characters* on a light background.

707.8 Braille Instructions. Braille instructions for initiating the speech mode shall be provided. Braille shall comply with 703.3.

708 Two-Way Communication Systems

708.1 General. Two-way communication systems shall comply with 708.

> **Advisory 708.1 General.** Devices that do not require handsets are easier to use by people who have a limited reach.

708.2 Audible and Visual Indicators. The system shall provide both audible and visual signals.

> **Advisory 708.2 Audible and Visual Indicators.** A light can be used to indicate visually that assistance is on the way. Signs indicating the meaning of visual signals should be provided.

708.3 Handsets. Handset cords, if provided, shall be 29 inches (735 mm) long minimum.

708.4 Residential Dwelling Unit Communication Systems. Communications systems between a *residential dwelling unit* and a *site*, *building*, or floor *entrance* shall comply with 708.4.

708.4.1 Common Use or Public Use System Interface. The *common use* or *public use* system interface shall include the capability of supporting voice and *TTY* communication with the *residential dwelling unit* interface.

708.4.2 Residential Dwelling Unit Interface. The *residential dwelling unit* system interface shall include a telephone jack capable of supporting voice and *TTY* communication with the *common use* or *public use* system interface.

CHAPTER 8: SPECIAL ROOMS, SPACES, AND ELEMENTS

801 General

801.1 Scope. The provisions of Chapter 8 shall apply where required by Chapter 2 or where referenced by a requirement in this document.

> **Advisory 801.1 Scope.** Facilities covered by these requirements are also subject to the requirements of the other chapters. For example, 806 addresses guest rooms in transient lodging facilities while 902 contains the technical specifications for dining surfaces. If a transient lodging facility contains a restaurant, the restaurant must comply with requirements in other chapters such as those applicable to certain dining surfaces.

802 Wheelchair Spaces, Companion Seats, and Designated Aisle Seats

802.1 Wheelchair Spaces. *Wheelchair spaces* shall comply with 802.1.

802.1.1 Floor or Ground Surface. The floor or ground surface of *wheelchair spaces* shall comply with 302. Changes in level are not permitted.
EXCEPTION: Slopes not steeper than 1:48 shall be permitted.

802.1.2 Width. A single *wheelchair space* shall be 36 inches (915 mm) wide minimum Where two adjacent *wheelchair spaces* are provided, each *wheelchair space* shall be 33 inches (840 mm) wide minimum.

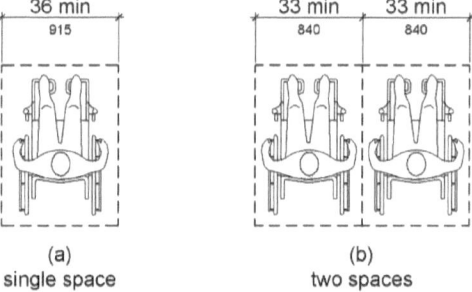

(a) single space (b) two spaces

Figure 802.1.2
Width of Wheelchair Spaces

802.1.3 Depth. Where a *wheelchair space* can be entered from the front or rear, the *wheelchair space* shall be 48 inches (1220 mm) deep minimum. Where a *wheelchair space* can be entered only from the side, the *wheelchair space* shall be 60 inches (1525 mm) deep minimum.

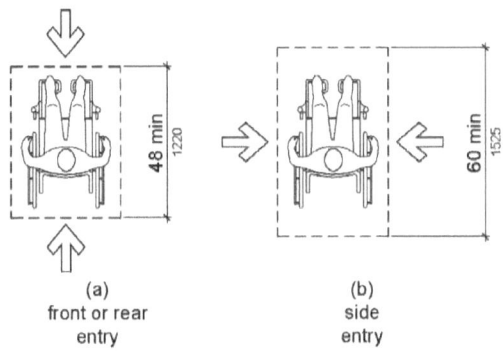

(a) front or rear entry
(b) side entry

Figure 802.1.3
Depth of Wheelchair Spaces

802.1.4 Approach. *Wheelchair spaces* shall adjoin *accessible* routes. *Accessible* routes shall not overlap *wheelchair spaces*.

> **Advisory 802.1.4 Approach.** Because accessible routes serving wheelchair spaces are not permitted to overlap the clear floor space at wheelchair spaces, access to any wheelchair space cannot be through another wheelchair space.

802.1.5 Overlap. *Wheelchair spaces* shall not overlap *circulation paths*.

> **Advisory 802.1.5 Overlap.** The term "circulation paths" used in Section 802.1.5 means aisle width required by applicable building or life safety codes for the specific assembly occupancy. Where the circulation path provided is wider than the required aisle width, the wheelchair space may intrude into that portion of the circulation path that is provided in excess of the required aisle width.

802.2 Lines of Sight. Lines of sight to the screen, performance area, or playing field for spectators in *wheelchair spaces* shall comply with 802.2.

802.2.1 Lines of Sight Over Seated Spectators. Where spectators are expected to remain seated during events, spectators in *wheelchair spaces* shall be afforded lines of sight complying with 802.2.1.

802.2.1.1 Lines of Sight Over Heads. Where spectators are provided lines of sight over the heads of spectators seated in the first row in front of their seats, spectators seated in *wheelchair spaces* shall be afforded lines of sight over the heads of seated spectators in the first row in front of *wheelchair spaces*.

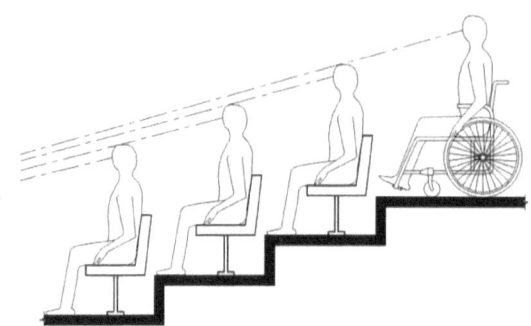

Figure 802.2.1.1
Lines of Sight Over the Heads of Seated Spectators

802.2.1.2 Lines of Sight Between Heads. Where spectators are provided lines of sight over the shoulders and between the heads of spectators seated in the first row in front of their seats, spectators seated in *wheelchair spaces* shall be afforded lines of sight over the shoulders and between the heads of seated spectators in the first row in front of *wheelchair spaces*.

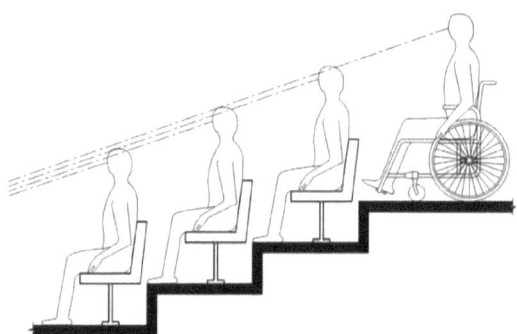

Figure 802.2.1.2
Lines of Sight Between the Heads of Seated Spectators

802.2.2 Lines of Sight Over Standing Spectators. Where spectators are expected to stand during events, spectators in *wheelchair spaces* shall be afforded lines of sight complying with 802.2.2.

802.2.2.1 Lines of Sight Over Heads. Where standing spectators are provided lines of sight over the heads of spectators standing in the first row in front of their seats, spectators seated in *wheelchair spaces* shall be afforded lines of sight over the heads of standing spectators in the first row in front of *wheelchair spaces*.

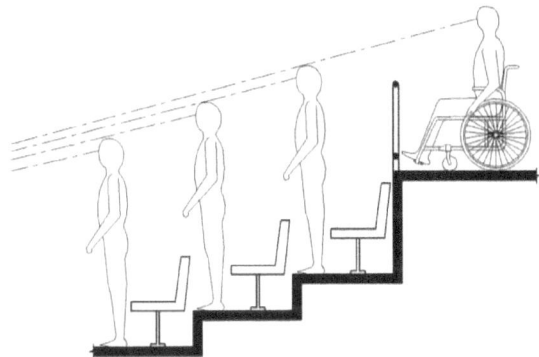

Figure 802.2.2.1
Lines of Sight Over the Heads of Standing Spectators

802.2.2.2 Lines of Sight Between Heads. Where standing spectators are provided lines of sight over the shoulders and between the heads of spectators standing in the first row in front of their seats, spectators seated in *wheelchair spaces* shall be afforded lines of sight over the shoulders and between the heads of standing spectators in the first row in front of *wheelchair spaces*.

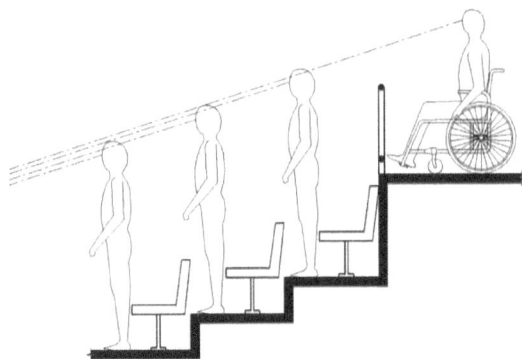

Figure 802.2.2.2
Lines of Sight Between the Heads of Standing Spectators

802.3 Companion Seats. Companion seats shall comply with 802.3.

802.3.1 Alignment. In row seating, companion seats shall be located to provide shoulder alignment with adjacent *wheelchair spaces*. The shoulder alignment point of the *wheelchair space* shall be measured 36 inches (915 mm) from the front of the *wheelchair space*. The floor surface of the companion seat shall be at the same elevation as the floor surface of the *wheelchair space*.

802.3.2 Type. Companion seats shall be equivalent in size, quality, comfort, and amenities to the seating in the immediate area. Companion seats shall be permitted to be movable.

802.4 Designated Aisle Seats. Designated aisle seats shall comply with 802.4.

802.4.1 Armrests. Where armrests are provided on the seating in the immediate area, folding or retractable armrests shall be provided on the aisle side of the seat.

802.4.2 Identification. Each designated aisle seat shall be identified by a sign or marker.

> **Advisory 802.4.2 Identification.** Seats with folding or retractable armrests are intended for use by individuals who have difficulty walking. Consider identifying such seats with signs that contrast (light-on-dark or dark-on-light) and that are also photo luminescent.

803 Dressing, Fitting, and Locker Rooms

803.1 General. Dressing, fitting, and locker rooms shall comply with 803.

> **Advisory 803.1 General.** Partitions and doors should be designed to ensure people using accessible dressing and fitting rooms privacy equivalent to that afforded other users of the facility. Section 903.5 requires dressing room bench seats to be installed so that they are at the same height as a typical wheelchair seat, 17 inches (430 mm) to 19 inches (485 mm). However, wheelchair seats can be lower than dressing room benches for people of short stature or children using wheelchairs.

803.2 Turning Space. Turning *space* complying with 304 shall be provided within the room.

803.3 Door Swing. Doors shall not swing into the room unless a clear floor or ground *space* complying with 305.3 is provided beyond the arc of the door swing.

803.4 Benches. A bench complying with 903 shall be provided within the room.

803.5 Coat Hooks and Shelves. Coat hooks provided within the room shall be located within one of the reach ranges specified in 308. Shelves shall be 40 inches (1015 mm) minimum and 48 inches (1220 mm) maximum above the finish floor or ground.

804 Kitchens and Kitchenettes

804.1 General. Kitchens and kitchenettes shall comply with 804.

804.2 Clearance. Where a pass through kitchen is provided, clearances shall comply with 804.2.1. Where a U-shaped kitchen is provided, clearances shall comply with 804.2.2.

EXCEPTION: *Spaces* that do not provide a cooktop or conventional range shall not be required to comply with 804.2.

> **Advisory 804.2 Clearance.** Clearances are measured from the furthest projecting face of all opposing base cabinets, counter tops, appliances, or walls, excluding hardware.

804.2.1 Pass Through Kitchen. In pass through kitchens where counters, appliances or cabinets are on two opposing sides, or where counters, appliances or cabinets are opposite a parallel wall, clearance between all opposing base cabinets, counter tops, appliances, or walls within kitchen work areas shall be 40 inches (1015 mm) minimum. Pass through kitchens shall have two entries.

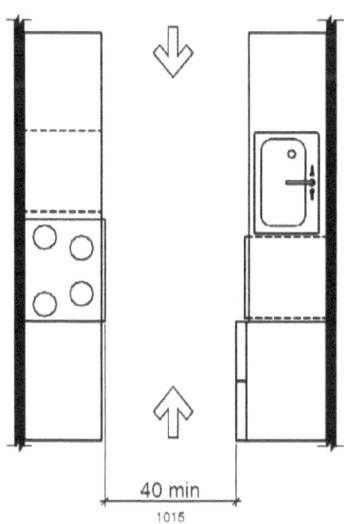

Figure 804.2.1
Pass Through Kitchens

804.2.2 U-Shaped. In U-shaped kitchens enclosed on three contiguous sides, clearance between all opposing base cabinets, counter tops, appliances, or walls within kitchen work areas shall be 60 inches (1525 mm) minimum.

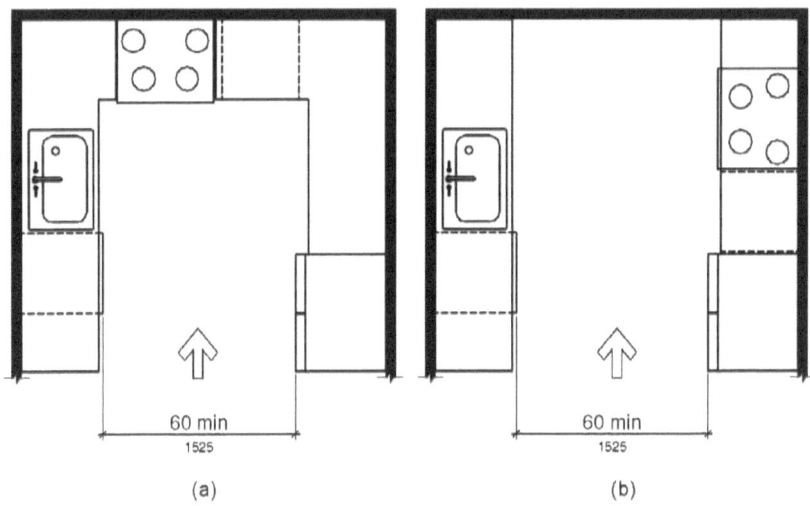

Figure 804.2.2
U-Shaped Kitchens

804.3 Kitchen Work Surface. In *residential dwelling units* required to comply with 809, at least one 30 inches (760 mm) wide minimum section of counter shall provide a kitchen work surface that complies with 804.3.

804.3.1 Clear Floor or Ground Space. A clear floor *space* complying with 305 positioned for a forward approach shall be provided. The clear floor or ground *space* shall be centered on the kitchen work surface and shall provide knee and toe clearance complying with 306.
    EXCEPTION: Cabinetry shall be permitted under the kitchen work surface provided that all of the following conditions are met:
(a) the cabinetry can be removed without removal or replacement of the kitchen work surface;
(b) the finish floor extends under the cabinetry; and
(c) the walls behind and surrounding the cabinetry are finished.

804.3.2 Height. The kitchen work surface shall be 34 inches (865 mm) maximum above the finish floor or ground.
    EXCEPTION: A counter that is adjustable to provide a kitchen work surface at variable heights, 29 inches (735 mm) minimum and 36 inches (915 mm) maximum, shall be permitted.

804.3.3 Exposed Surfaces. There shall be no sharp or abrasive surfaces under the work surface counters.

CHAPTER 8: SPECIAL ROOMS, SPACES, AND ELEMENTS   TECHNICAL

804.4 Sinks. Sinks shall comply with 606.

804.5 Storage. At least 50 percent of shelf *space* in storage *facilities* shall comply with 811.

804.6 Appliances. Where provided, kitchen appliances shall comply with 804.6.

>   **804.6.1 Clear Floor or Ground Space.** A clear floor or ground *space* complying with 305 shall be provided at each kitchen appliance. Clear floor or ground *spaces* shall be permitted to overlap.
>
>   **804.6.2 Operable Parts.** All appliance controls shall comply with 309.
>   **EXCEPTIONS: 1.** Appliance doors and door latching devices shall not be required to comply with 309.4.
>   **2.** Bottom-hinged appliance doors, when in the open position, shall not be required to comply with 309.3.
>
>   **804.6.3 Dishwasher.** Clear floor or ground *space* shall be positioned adjacent to the dishwasher door. The dishwasher door, in the open position, shall not obstruct the clear floor or ground *space* for the dishwasher or the sink.
>
>   **804.6.4 Range or Cooktop.** Where a forward approach is provided, the clear floor or ground *space* shall provide knee and toe clearance complying with 306. Where knee and toe *space* is provided, the underside of the range or cooktop shall be insulated or otherwise configured to prevent burns, abrasions, or electrical shock. The location of controls shall not require reaching across burners.
>
>   **804.6.5 Oven.** Ovens shall comply with 804.6.5.
>
>>   **804.6.5.1 Side-Hinged Door Ovens.** Side-hinged door ovens shall have the work surface required by 804.3 positioned adjacent to the latch side of the oven door.
>>
>>   **804.6.5.2 Bottom-Hinged Door Ovens.** Bottom-hinged door ovens shall have the work surface required by 804.3 positioned adjacent to one side of the door.
>>
>>   **804.6.5.3 Controls.** Ovens shall have controls on front panels.
>
>   **804.6.6 Refrigerator/Freezer.** Combination refrigerators and freezers shall have at least 50 percent of the freezer *space* 54 inches (1370 mm) maximum above the finish floor or ground. The clear floor or ground *space* shall be positioned for a parallel approach to the *space* dedicated to a refrigerator/freezer with the centerline of the clear floor or ground *space* offset 24 inches (610 mm) maximum from the centerline of the dedicated *space*.

805 Medical Care and Long-Term Care Facilities

805.1 General. Medical care *facility* and long-term care *facility* patient or resident sleeping rooms required to provide mobility features shall comply with 805.

805.2 Turning Space. Turning *space* complying with 304 shall be provided within the room.

805.3 Clear Floor or Ground Space. A clear floor *space* complying with 305 shall be provided on each side of the bed. The clear floor *space* shall be positioned for parallel approach to the side of the bed.

805.4 Toilet and Bathing Rooms. Toilet and bathing rooms that are provided as part of a patient or resident sleeping room shall comply with 603. Where provided, no fewer than one water closet, one lavatory, and one bathtub or shower shall comply with the applicable requirements of 603 through 610.

806 Transient Lodging Guest Rooms

806.1 General. *Transient lodging* guest rooms shall comply with 806. Guest rooms required to provide mobility features shall comply with 806.2. Guest rooms required to provide communication features shall comply with 806.3.

806.2 Guest Rooms with Mobility Features. Guest rooms required to provide mobility features shall comply with 806.2.

> **Advisory 806.2 Guest Rooms.** The requirements in Section 806.2 do not include requirements that are common to all accessible spaces. For example, closets in guest rooms must comply with the applicable provisions for storage specified in scoping.

806.2.1 Living and Dining Areas. Living and dining areas shall be *accessible*.

806.2.2 Exterior Spaces. Exterior *spaces*, including patios, terraces and balconies, that serve the guest room shall be *accessible*.

806.2.3 Sleeping Areas. At least one sleeping area shall provide a clear floor *space* complying with 305 on both sides of a bed. The clear floor *space* shall be positioned for parallel approach to the side of the bed.
> **EXCEPTION:** Where a single clear floor *space* complying with 305 positioned for parallel approach is provided between two beds, a clear floor or ground *space* shall not be required on both sides of a bed.

806.2.4 Toilet and Bathing Facilities. At least one bathroom that is provided as part of a guest room shall comply with 603. No fewer than one water closet, one lavatory, and one bathtub or shower shall comply with applicable requirements of 603 through 610. In addition, required roll-in shower compartments shall comply with 608.2.2 or 608.2.3. Toilet and bathing fixtures required to comply with 603 through 610 shall be permitted to be located in more than one toilet or bathing area, provided that travel between fixtures does not require travel between other parts of the guest room.

806.2.4.1 Vanity Counter Top Space. If vanity counter top *space* is provided in non-accessible guest toilet or bathing rooms, comparable vanity counter top *space*, in terms of size and proximity to the lavatory, shall also be provided in *accessible* guest toilet or bathing rooms.

> **Advisory 806.2.4.1 Vanity Counter Top Space.** This provision is intended to ensure that accessible guest rooms are provided with comparable vanity counter top space.

806.2.5 Kitchens and Kitchenettes. Kitchens and kitchenettes shall comply with 804.

806.2.6 Turning Space. Turning *space* complying with 304 shall be provided within the guest room.

806.3 Guest Rooms with Communication Features. Guest rooms required to provide communication features shall comply with 806.3.

> **Advisory 806.3 Guest Rooms with Communication Features.** In guest rooms required to have accessible communication features, consider ensuring compatibility with adaptive equipment used by people with hearing impairments. To ensure communication within the facility, as well as on commercial lines, provide telephone interface jacks that are compatible with both digital and analog signal use. If an audio headphone jack is provided on a speaker phone, a cutoff switch can be included in the jack so that insertion of the jack cuts off the speaker. If a telephone-like handset is used, the external speakers can be turned off when the handset is removed from the cradle. For headset or external amplification system compatibility, a standard subminiature jack installed in the telephone will provide the most flexibility.

806.3.1 Alarms. Where emergency warning systems are provided, alarms complying with 702 shall be provided.

806.3.2 Notification Devices. Visible notification devices shall be provided to alert room occupants of incoming telephone calls and a door knock or bell. Notification devices shall not be connected to visible alarm signal appliances. Telephones shall have volume controls compatible with the telephone system and shall comply with 704.3. Telephones shall be served by an electrical outlet complying with 309 located within 48 inches (1220 mm) of the telephone to facilitate the use of a *TTY*.

807 Holding Cells and Housing Cells

807.1 General. Holding cells and housing cells shall comply with 807.

807.2 Cells with Mobility Features. Cells required to provide mobility features shall comply with 807.2.

807.2.1 Turning Space. Turning *space* complying with 304 shall be provided within the cell.

807.2.2 Benches. Where benches are provided, at least one bench shall comply with 903.

807.2.3 Beds. Where beds are provided, clear floor *space* complying with 305 shall be provided on at least one side of the bed. The clear floor *space* shall be positioned for parallel approach to the side of the bed.

807.2.4 Toilet and Bathing Facilities. Toilet *facilities* or bathing *facilities* that are provided as part of a cell shall comply with 603. Where provided, no fewer than one water closet, one lavatory, and one bathtub or shower shall comply with the applicable requirements of 603 through 610.

> **Advisory 807.2.4 Toilet and Bathing Facilities.** In holding cells, housing cells, or rooms required to be accessible, these requirements do not require a separate toilet room.

807.3 Cells with Communication Features. Cells required to provide communication features shall comply with 807.3.

807.3.1 Alarms. Where audible emergency alarm systems are provided to serve the occupants of cells, visible alarms complying with 702 shall be provided.

EXCEPTION: Visible alarms shall not be required where inmates or detainees are not allowed independent means of egress.

807.3.2 Telephones. Telephones, where provided within cells, shall have volume controls complying with 704.3.

808 Courtrooms

808.1 General. Courtrooms shall comply with 808.

808.2 Turning Space. Where provided, areas that are raised or depressed and accessed by *ramps* or platform lifts with entry *ramps* shall provide unobstructed turning *space* complying with 304.

808.3 Clear Floor Space. Each jury box and witness stand shall have, within its defined area, clear floor *space* complying with 305.

EXCEPTION: In *alterations*, *wheelchair spaces* are not required to be located within the defined area of raised jury boxes or witness stands and shall be permitted to be located outside these *spaces* where *ramp* or platform lift access poses a hazard by restricting or projecting into a means of egress required by the appropriate *administrative authority*.

808.4 Judges' Benches and Courtroom Stations. Judges' benches, clerks' stations, bailiffs' stations, deputy clerks' stations, court reporters' stations and litigants' and counsel stations shall comply with 902.

809 Residential Dwelling Units

809.1 General. *Residential dwelling units* shall comply with 809. *Residential dwelling units* required to provide mobility features shall comply with 809.2 through 809.4. *Residential dwelling units* required to provide communication features shall comply with 809.5.

809.2 Accessible Routes. *Accessible* routes complying with Chapter 4 shall be provided within *residential dwelling units* in accordance with 809.2.

EXCEPTION: *Accessible* routes shall not be required to or within unfinished attics or unfinished basements.

809.2.1 Location. At least one *accessible* route shall connect all *spaces* and *elements* which are a part of the *residential dwelling unit*. Where only one *accessible* route is provided, it shall not pass through bathrooms, closets, or similar *spaces*.

809.2.2 Turning Space. All rooms served by an *accessible* route shall provide a turning *space* complying with 304.
   EXCEPTION: Turning *space* shall not be required in exterior *spaces* 30 inches (760 mm) maximum in depth or width.

> **Advisory 809.2.2 Turning Space.** It is generally acceptable to use required clearances to provide wheelchair turning space. For example, in kitchens, 804.3.1 requires at least one work surface with clear floor space complying with 306 to be centered beneath. If designers elect to provide clear floor space that is at least 36 inches (915 mm) wide, as opposed to the required 30 inches (760 mm) wide, that clearance can be part of a T-turn, thereby maximizing efficient use of the kitchen area. However, the overlap of turning space must be limited to one segment of the T-turn so that back-up maneuvering is not restricted. It would, therefore, be unacceptable to use both the clearances under the work surface and the sink as part of a T-turn. See Section 304.3.2 regarding T-turns.

809.3 Kitchen. Where a kitchen is provided, it shall comply with 804.

809.4 Toilet Facilities and Bathing Facilities. At least one bathroom shall comply with 603. No fewer than one of each type of fixture provided shall comply with applicable requirements of 603 through 610. Toilet and bathing fixtures required to comply with 603 through 610 shall be located in the same toilet and bathing area, such that travel between fixtures does not require travel between other parts of the *residential dwelling unit*.

> **Advisory 809.4 Toilet Facilities and Bathing Facilities.** In an effort to promote space efficiency, vanity counter top space in accessible residential dwelling units is often omitted. This omission does not promote equal access or equal enjoyment of the unit. Where comparable units have vanity counter tops, accessible units should also have vanity counter tops located as close as possible to the lavatory for convenient access to toiletries.

809.5 Residential Dwelling Units with Communication Features. *Residential dwelling units* required to provide communication features shall comply with 809.5.

   809.5.1 Building Fire Alarm System. Where a *building* fire alarm system is provided, the system wiring shall be extended to a point within the *residential dwelling unit* in the vicinity of the *residential dwelling unit* smoke detection system.

      809.5.1.1 Alarm Appliances. Where alarm appliances are provided within a *residential dwelling unit* as part of the *building* fire alarm system, they shall comply with 702.

      809.5.1.2 Activation. All visible alarm appliances provided within the *residential dwelling unit* for *building* fire alarm notification shall be activated upon activation of the *building* fire alarm in the portion of the *building* containing the *residential dwelling unit*.

   809.5.2 Residential Dwelling Unit Smoke Detection System. *Residential dwelling unit* smoke detection systems shall comply with NFPA 72 (1999 or 2002 edition) (incorporated by reference, see "Referenced Standards" in Chapter 1).

809.5.2.1 Activation. All visible alarm appliances provided within the *residential dwelling unit* for smoke detection notification shall be activated upon smoke detection.

809.5.3 Interconnection. The same visible alarm appliances shall be permitted to provide notification of *residential dwelling unit* smoke detection and *building* fire alarm activation.

809.5.4 Prohibited Use. Visible alarm appliances used to indicate *residential dwelling unit* smoke detection or *building* fire alarm activation shall not be used for any other purpose within the *residential dwelling unit*.

809.5.5 Residential Dwelling Unit Primary Entrance. Communication features shall be provided at the *residential dwelling unit* primary *entrance* complying with 809.5.5.

809.5.5.1 Notification. A hard-wired electric doorbell shall be provided. A button or switch shall be provided outside the *residential dwelling unit* primary *entrance*. Activation of the button or switch shall initiate an audible tone and visible signal within the *residential dwelling unit*. Where visible doorbell signals are located in sleeping areas, they shall have controls to deactivate the signal.

809.5.5.2 Identification. A means for visually identifying a visitor without opening the *residential dwelling unit* entry door shall be provided and shall allow for a minimum 180 degree range of view.

> **Advisory 809.5.5.2 Identification.** In doors, peepholes that include prisms clarify the image and should offer a wide-angle view of the hallway or exterior for both standing persons and wheelchair users. Such peepholes can be placed at a standard height and permit a view from several feet from the door.

809.5.6 Site, Building, or Floor Entrance. Where a system, including a closed-circuit system, permitting voice communication between a visitor and the occupant of the *residential dwelling unit* is provided, the system shall comply with 708.4.

810 Transportation Facilities

810.1 General. Transportation *facilities* shall comply with 810.

810.2 Bus Boarding and Alighting Areas. Bus boarding and alighting areas shall comply with 810.2.

> **Advisory 810.2 Bus Boarding and Alighting Areas.** At bus stops where a shelter is provided, the bus stop pad can be located either within or outside of the shelter.

810.2.1 Surface. Bus stop boarding and alighting areas shall have a firm, stable surface.

810.2.2 Dimensions. Bus stop boarding and alighting areas shall provide a clear length of 96 inches (2440 mm) minimum, measured perpendicular to the curb or vehicle roadway edge, and a clear width of 60 inches (1525 mm) minimum, measured parallel to the vehicle roadway.

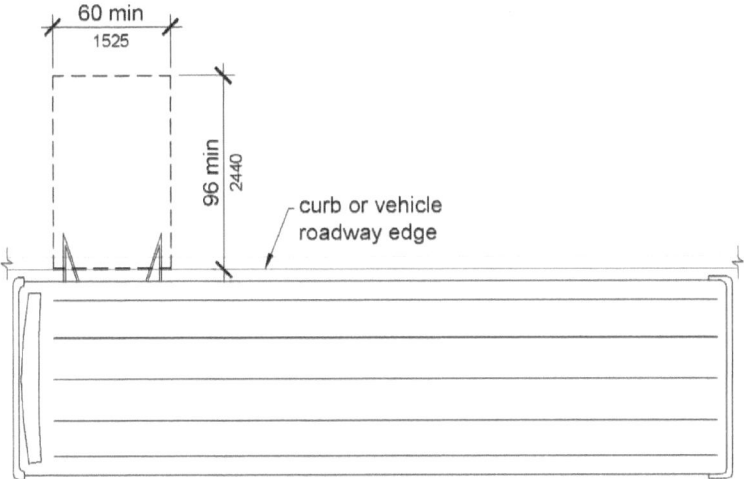

Figure 810.2.2
Dimensions of Bus Boarding and Alighting Areas

810.2.3 Connection. Bus stop boarding and alighting areas shall be connected to streets, sidewalks, or pedestrian paths by an *accessible* route complying with 402.

810.2.4 Slope. Parallel to the roadway, the slope of the bus stop boarding and alighting area shall be the same as the roadway, to the maximum extent practicable. Perpendicular to the roadway, the slope of the bus stop boarding and alighting area shall not be steeper than 1:48.

810.3 Bus Shelters. Bus shelters shall provide a minimum clear floor or ground *space* complying with 305 entirely within the shelter. Bus shelters shall be connected by an *accessible* route complying with 402 to a boarding and alighting area complying with 810.2.

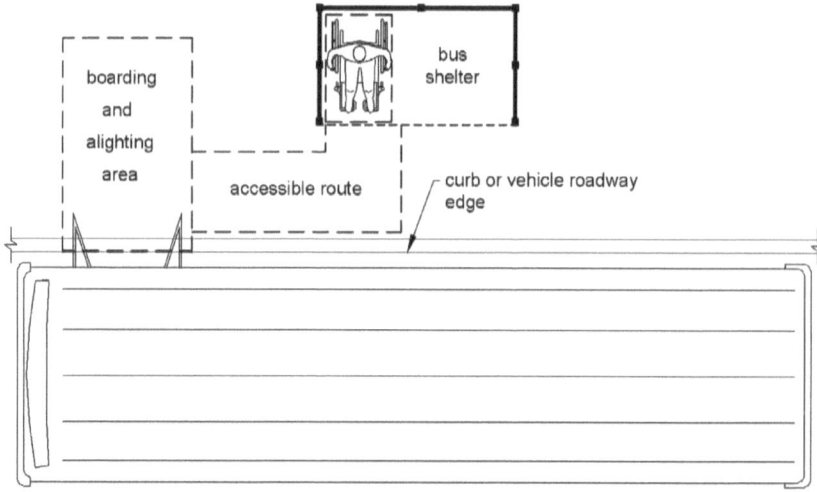

**Figure 810.3
Bus Shelters**

810.4 Bus Signs. Bus route identification signs shall comply with 703.5.1 through 703.5.4, and 703.5.7 and 703.5.8. In addition, to the maximum extent practicable, bus route identification signs shall comply with 703.5.5.

    EXCEPTION: Bus schedules, timetables and maps that are posted at the bus stop or bus bay shall not be required to comply.

810.5 Rail Platforms. Rail platforms shall comply with 810.5.

    810.5.1 Slope. Rail platforms shall not exceed a slope of 1:48 in all directions.

        EXCEPTION: Where platforms serve vehicles operating on existing track or track laid in existing roadway, the slope of the platform parallel to the track shall be permitted to be equal to the slope (grade) of the roadway or existing track.

    810.5.2 Detectable Warnings. Platform boarding edges not protected by platform screens or guards shall have *detectable warnings* complying with 705 along the full length of the *public use* area of the platform.

    810.5.3 Platform and Vehicle Floor Coordination. Station platforms shall be positioned to coordinate with vehicles in accordance with the applicable requirements of 36 CFR Part 1192. Low-level platforms shall be 8 inches (205 mm) minimum above top of rail.

EXCEPTION: Where vehicles are boarded from sidewalks or street-level, low-level platforms shall be permitted to be less than 8 inches (205 mm).

> **Advisory 810.5.3 Platform and Vehicle Floor Coordination.** The height and position of a platform must be coordinated with the floor of the vehicles it serves to minimize the vertical and horizontal gaps, in accordance with the ADA Accessibility Guidelines for Transportation Vehicles (36 CFR Part 1192). The vehicle guidelines, divided by bus, van, light rail, rapid rail, commuter rail, intercity rail, are available at www.access-board.gov. The preferred alignment is a high platform, level with the vehicle floor. In some cases, the vehicle guidelines permit use of a low platform in conjunction with a lift or ramp. Most such low platforms must have a minimum height of eight inches above the top of the rail. Some vehicles are designed to be boarded from a street or the sidewalk along the street and the exception permits such boarding areas to be less than eight inches high.

810.6 Rail Station Signs. Rail station signs shall comply with 810.6.
EXCEPTION. Signs shall not be required to comply with 810.6.1 and 810.6.2 where audible signs are remotely transmitted to hand-held receivers, or are user- or proximity-actuated.

> **Advisory 810.6 Rail Station Signs Exception.** Emerging technologies such as an audible sign systems using infrared transmitters and receivers may provide greater accessibility in the transit environment than traditional Braille and raised letter signs. The transmitters are placed on or next to print signs and transmit their information to an infrared receiver that is held by a person. By scanning an area, the person will hear the sign. This means that signs can be placed well out of reach of Braille readers, even on parapet walls and on walls beyond barriers. Additionally, such signs can be used to provide wayfinding information that cannot be efficiently conveyed on Braille signs.

810.6.1 Entrances. Where signs identify a station or its *entrance*, at least one sign at each *entrance* shall comply with 703.2 and shall be placed in uniform locations to the maximum extent practicable. Where signs identify a station that has no defined *entrance*, at least one sign shall comply with 703.2 and shall be placed in a central location.

810.6.2 Routes and Destinations. Lists of stations, routes and destinations served by the station which are located on boarding areas, platforms, or *mezzanines* shall comply with 703.5. At least one *tactile* sign identifying the specific station and complying with 703.2 shall be provided on each platform or boarding area. Signs covered by this requirement shall, to the maximum extent practicable, be placed in uniform locations within the system.
EXCEPTION: Where sign *space* is limited, *characters* shall not be required to exceed 3 inches (75 mm).

> **Advisory 810.6.2 Routes and Destinations.** Route maps are not required to comply with the informational sign requirements in this document.

810.6.3 Station Names. Stations covered by this section shall have identification signs complying with 703.5. Signs shall be clearly visible and within the sight lines of standing and sitting passengers from within the vehicle on both sides when not obstructed by another vehicle.

> **Advisory 810.6.3 Station Names.** It is also important to place signs at intervals in the station where passengers in the vehicle will be able to see a sign when the vehicle is either stopped at the station or about to come to a stop in the station. The number of signs necessary may be directly related to the size of the lettering displayed on the sign.

810.7 Public Address Systems. Where public address systems convey audible information to the public, the same or equivalent information shall be provided in a visual format.

810.8 Clocks. Where clocks are provided for use by the public, the clock face shall be uncluttered so that its *elements* are clearly visible. Hands, numerals and digits shall contrast with the background either light-on-dark or dark-on-light. Where clocks are installed overhead, numerals and digits shall comply with 703.5.

810.9 Escalators. Where provided, escalators shall comply with the sections 6.1.3.5.6 and 6.1.3.6.5 of ASME A17.1 (incorporated by reference, see "Referenced Standards" in Chapter 1) and shall have a clear width of 32 inches (815 mm) minimum.
   EXCEPTION: Existing escalators in *key stations* shall not be required to comply with 810.9.

810.10 Track Crossings. Where a *circulation path* serving boarding platforms crosses tracks, it shall comply with 402.
   EXCEPTION: Openings for wheel flanges shall be permitted to be 2½ inches (64 mm) maximum.

Figure 810.10 (Exception)
Track Crossings

811 Storage

811.1 General. Storage shall comply with 811.

811.2 Clear Floor or Ground Space. A clear floor or ground *space* complying with 305 shall be provided.

811.3 Height. Storage *elements* shall comply with at least one of the reach ranges specified in 308.

811.4 Operable Parts. *Operable parts* shall comply with 309.

CHAPTER 9: BUILT-IN ELEMENTS

901 General

901.1 Scope. The provisions of Chapter 9 shall apply where required by Chapter 2 or where referenced by a requirement in this document.

902 Dining Surfaces and Work Surfaces

902.1 General. Dining surfaces and work surfaces shall comply with 902.2 and 902.3.
   EXCEPTION: Dining surfaces and work surfaces for *children's use* shall be permitted to comply with 902.4.

> **Advisory 902.1 General.** Dining surfaces include, but are not limited to, bars, tables, lunch counters, and booths. Examples of work surfaces include writing surfaces, study carrels, student laboratory stations, baby changing and other tables or fixtures for personal grooming, coupon counters, and where covered by the ABA scoping provisions, employee work stations.

902.2 Clear Floor or Ground Space. A clear floor *space* complying with 305 positioned for a forward approach shall be provided. Knee and toe clearance complying with 306 shall be provided.

902.3 Height. The tops of dining surfaces and work surfaces shall be 28 inches (710 mm) minimum and 34 inches (865 mm) maximum above the finish floor or ground.

902.4 Dining Surfaces and Work Surfaces for Children's Use. *Accessible* dining surfaces and work surfaces for *children's use* shall comply with 902.4.
   EXCEPTION: Dining surfaces and work surfaces that are used primarily by children 5 years and younger shall not be required to comply with 902.4 where a clear floor or ground *space* complying with 305 positioned for a parallel approach is provided.

   902.4.1 Clear Floor or Ground Space. A clear floor *space* complying with 305 positioned for forward approach shall be provided. Knee and toe clearance complying with 306 shall be provided, except that knee clearance 24 inches (610 mm) minimum above the finish floor or ground shall be permitted.

   902.4.2 Height. The tops of tables and counters shall be 26 inches (660 mm) minimum and 30 inches (760 mm) maximum above the finish floor or ground.

903 Benches

903.1 General. Benches shall comply with 903.

903.2 Clear Floor or Ground Space. Clear floor or ground *space* complying with 305 shall be provided and shall be positioned at the end of the bench seat and parallel to the short axis of the bench.

| TECHNICAL | CHAPTER 9: BUILT-IN ELEMENTS |

903.3 Size. Benches shall have seats that are 42 inches (1065 mm) long minimum and 20 inches (510 mm) deep minimum and 24 inches (610 mm) deep maximum.

903.4 Back Support. The bench shall provide for back support or shall be affixed to a wall. Back support shall be 42 inches (1065 mm) long minimum and shall extend from a point 2 inches (51 mm) maximum above the seat surface to a point 18 inches (455 mm) minimum above the seat surface. Back support shall be 2½ inches (64 mm) maximum from the rear edge of the seat measured horizontally.

> **Advisory 903.4 Back Support.** To assist in transferring to the bench, consider providing grab bars on a wall adjacent to the bench, but not on the seat back. If provided, grab bars cannot obstruct transfer to the bench.

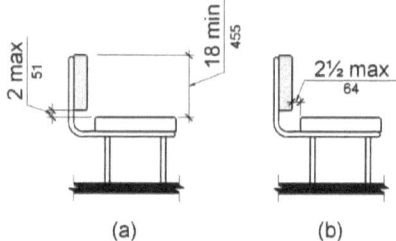

Figure 903.4
Bench Back Support

903.5 Height. The top of the bench seat surface shall be 17 inches (430 mm) minimum and 19 inches (485 mm) maximum above the finish floor or ground.

903.6 Structural Strength. Allowable stresses shall not be exceeded for materials used when a vertical or horizontal force of 250 pounds (1112 N) is applied at any point on the seat, fastener, mounting device, or supporting structure.

903.7 Wet Locations. Where installed in wet locations, the surface of the seat shall be slip resistant and shall not accumulate water.

904 Check-Out Aisles and Sales and Service Counters

904.1 General. Check-out aisles and sales and service counters shall comply with the applicable requirements of 904.

904.2 Approach. All portions of counters required to comply with 904 shall be located adjacent to a walking surface complying with 403.

> **Advisory 904.2 Approach.** If a cash register is provided at the sales or service counter, locate the accessible counter close to the cash register so that a person using a wheelchair is visible to sales or service personnel and to minimize the reach for a person with a disability.

904.3 Check-Out Aisles. Check-out aisles shall comply with 904.3.

904.3.1 Aisle. Aisles shall comply with 403.

904.3.2 Counter. The counter surface height shall be 38 inches (965 mm) maximum above the finish floor or ground. The top of the counter edge protection shall be 2 inches (51 mm) maximum above the top of the counter surface on the aisle side of the check-out counter.

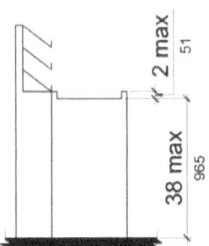

Figure 904.3.2
Check-Out Aisle Counters

904.3.3 Check Writing Surfaces. Where provided, check writing surfaces shall comply with 902.3.

904.4 Sales and Service Counters. Sales counters and service counters shall comply with 904.4.1 or 904.4.2. The *accessible* portion of the counter top shall extend the same depth as the sales or service counter top.
   EXCEPTION: In *alterations*, when the provision of a counter complying with 904.4 would result in a reduction of the number of existing counters at work stations or a reduction of the number of existing *mail boxes*, the counter shall be permitted to have a portion which is 24 inches (610 mm) long minimum complying with 904.4.1 provided that the required clear floor or ground *space* is centered on the *accessible* length of the counter.

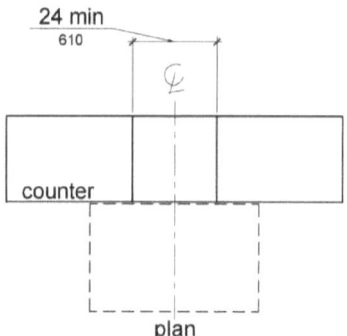

Figure 904.4 (Exception)
Alteration of Sales and Service Counters

904.4.1 Parallel Approach. A portion of the counter surface that is 36 inches (915 mm) long minimum and 36 inches (915 mm) high maximum above the finish floor shall be provided. A clear floor or ground *space* complying with 305 shall be positioned for a parallel approach adjacent to the 36 inch (915 mm) minimum length of counter.

    EXCEPTION: Where the provided counter surface is less than 36 inches (915 mm) long, the entire counter surface shall be 36 inches (915 mm) high maximum above the finish floor.

904.4.2 Forward Approach. A portion of the counter surface that is 30 inches (760 mm) long minimum and 36 inches (915 mm) high maximum shall be provided. Knee and toe *space* complying with 306 shall be provided under the counter. A clear floor or ground *space* complying with 305 shall be positioned for a forward approach to the counter.

904.5 Food Service Lines. Counters in food service lines shall comply with 904.5.

    904.5.1 Self-Service Shelves and Dispensing Devices. Self-service shelves and dispensing devices for tableware, dishware, condiments, food and beverages shall comply with 308.

    904.5.2 Tray Slides. The tops of tray slides shall be 28 inches (710 mm) minimum and 34 inches (865 mm) maximum above the finish floor or ground.

904.6 Security Glazing. Where counters or teller windows have security glazing to separate personnel from the public, a method to facilitate voice communication shall be provided. Telephone handset devices, if provided, shall comply with 704.3.

Advisory 904.6 Security Glazing. Assistive listening devices complying with 706 can facilitate voice communication at counters or teller windows where there is security glazing which promotes distortion in audible information. Where assistive listening devices are installed, place signs complying with 703.7.2.4 to identify those facilities which are so equipped. Other voice communication methods include, but are not limited to, grilles, slats, talk-through baffles, intercoms, or telephone handset devices.

CHAPTER 10: RECREATION FACILITIES

1001 General

1001.1 Scope. The provisions of Chapter 10 shall apply where required by Chapter 2 or where referenced by a requirement in this document.

> **Advisory 1001.1 Scope.** Unless otherwise modified or specifically addressed in Chapter 10, all other ADAAG provisions apply to the design and construction of recreation facilities and elements. The provisions in Section 1001.1 apply wherever these elements are provided. For example, office buildings may contain a room with exercise equipment to which these sections would apply.

1002 Amusement Rides

1002.1 General. *Amusement rides* shall comply with 1002.

1002.2 Accessible Routes. *Accessible* routes serving *amusement rides* shall comply with Chapter 4.
EXCEPTIONS: 1. In load or unload areas and on *amusement rides*, where compliance with 405.2 is not structurally or operationally feasible, *ramp* slope shall be permitted to be 1:8 maximum.
2. In load or unload areas and on *amusement rides*, handrails provided along walking surfaces complying with 403 and required on *ramps* complying with 405 shall not be required to comply with 505 where compliance is not structurally or operationally feasible.

> **Advisory 1002.2 Accessible Routes Exception 1.** Steeper slopes are permitted on accessible routes connecting the amusement ride in the load and unload position where it is "structurally or operationally infeasible." In most cases, this will be limited to areas where the accessible route leads directly to the amusement ride and where there are space limitations on the ride, not the queue line. Where possible, the least possible slope should be used on the accessible route that serves the amusement ride.

1002.3 Load and Unload Areas. A turning *space* complying with 304.2 and 304.3 shall be provided in load and unload areas.

1002.4 Wheelchair Spaces in Amusement Rides. *Wheelchair spaces* in *amusement rides* shall comply with 1002.4.

1002.4.1 Floor or Ground Surface. The floor or ground surface of *wheelchair spaces* shall be stable and firm.

1002.4.2 Slope. The floor or ground surface of *wheelchair spaces* shall have a slope not steeper than 1:48 when in the load and unload position.

1002.4.3 Gaps. Floors of *amusement rides* with *wheelchair spaces* and floors of load and unload areas shall be coordinated so that, when *amusement rides* are at rest in the load and unload

position, the vertical difference between the floors shall be within plus or minus 5/8 inches (16 mm) and the horizontal gap shall be 3 inches (75 mm) maximum under normal passenger load conditions.
  EXCEPTION: Where compliance is not operationally or structurally feasible, *ramps*, bridge plates, or similar devices complying with the applicable requirements of 36 CFR 1192.83(c) shall be provided.

> **Advisory 1002.4.3 Gaps Exception.** 36 CFR 1192.83(c) ADA Accessibility Guidelines for Transportation Vehicles - Light Rail Vehicles and Systems - Mobility Aid Accessibility is available at www.access-board.gov. It includes provisions for bridge plates and ramps that can be used at gaps between wheelchair spaces and floors of load and unload areas.

1002.4.4 Clearances. Clearances for *wheelchair spaces* shall comply with 1002.4.4.
  EXCEPTIONS: 1. Where provided, securement devices shall be permitted to overlap required clearances.
  2. *Wheelchair spaces* shall be permitted to be mechanically or manually repositioned.
  3. *Wheelchair spaces* shall not be required to comply with 307.4.

> **Advisory 1002.4.4 Clearances Exception 3.** This exception for protruding objects applies to the ride devices, not to circulation areas or accessible routes in the queue lines or the load and unload areas.

1002.4.4.1 Width and Length. *Wheelchair spaces* shall provide a clear width of 30 inches (760 mm) minimum and a clear length of 48 inches (1220 mm) minimum measured to 9 inches (230 mm) minimum above the floor surface.

1002.4.4.2 Side Entry. Where *wheelchair spaces* are entered only from the side, *amusement rides* shall be designed to permit sufficient maneuvering clearance for individuals using a wheelchair or mobility aid to enter and exit the ride.

> **Advisory 1002.4.4.2 Side Entry.** The amount of clear space needed within the ride, and the size and position of the opening are interrelated. A 32 inch (815 mm) clear opening will not provide sufficient width when entered through a turn into an amusement ride. Additional space for maneuvering and a wider door will be needed where a side opening is centered on the ride. For example, where a 42 inch (1065 mm) opening is provided, a minimum clear space of 60 inches (1525 mm) in length and 36 inches (915mm) in depth is needed to ensure adequate space for maneuvering.

1002.4.4.3 Permitted Protrusions in Wheelchair Spaces. Objects are permitted to protrude a distance of 6 inches (150 mm) maximum along the front of the *wheelchair space*, where located 9 inches (230 mm) minimum and 27 inches (685 mm) maximum above the floor or ground surface of the *wheelchair space*. Objects are permitted to protrude a distance of 25 inches (635 mm) maximum along the front of the *wheelchair space*, where located more than 27 inches (685 mm) above the floor or ground surface of the *wheelchair space*.

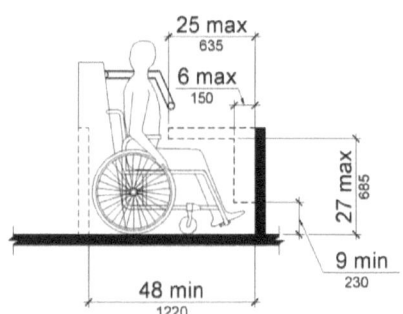

Figure 1002.4.4.3
Protrusions in Wheelchair Spaces in Amusement Rides

1002.4.5 Ride Entry. Openings providing entry to *wheelchair spaces* on *amusement rides* shall be 32 inches (815 mm) minimum clear.

1002.4.6 Approach. One side of the *wheelchair space* shall adjoin an *accessible* route when in the load and unload position.

1002.4.7 Companion Seats. Where the interior width of the *amusement ride* is greater than 53 inches (1345 mm), seating is provided for more than one rider, and the wheelchair is not required to be centered within the *amusement ride*, a companion seat shall be provided for each *wheelchair space*.

>**1002.4.7.1 Shoulder-to-Shoulder Seating.** Where an *amusement ride* provides shoulder-to-shoulder seating, companion seats shall be shoulder-to-shoulder with the adjacent *wheelchair space*.
>**EXCEPTION:** Where shoulder-to-shoulder companion seating is not operationally or structurally feasible, compliance with this requirement shall be required to the maximum extent practicable.

1002.5 Amusement Ride Seats Designed for Transfer. *Amusement ride seats* designed for transfer shall comply with 1002.5 when positioned for loading and unloading.

CHAPTER 10: RECREATION FACILITIES   TECHNICAL

> **Advisory 1002.5 Amusement Ride Seats Designed for Transfer.** The proximity of the clear floor or ground space next to an element and the height of the element one is transferring to are both critical for a safe and independent transfer. Providing additional clear floor or ground space both in front of and diagonal to the element will provide flexibility and will increase usability for a more diverse population of individuals with disabilities. Ride seats designed for transfer should involve only one transfer. Where possible, designers are encouraged to locate the ride seat no higher than 17 to 19 inches (430 to 485 mm) above the load and unload surface. Where greater distances are required for transfers, providing gripping surfaces, seat padding, and avoiding sharp objects in the path of transfer will facilitate the transfer.

1002.5.1 Clear Floor or Ground Space. A clear floor or ground *space* complying with 305 shall be provided in the load and unload area adjacent to the *amusement ride seats* designed for transfer.

1002.5.2 Transfer Height. The height of *amusement ride seats* designed for transfer shall be 14 inches (355 mm) minimum and 24 inches (610 mm) maximum measured from the surface of the load and unload area.

1002.5.3 Transfer Entry. Where openings are provided for transfer to *amusement ride seats*, the openings shall provide clearance for transfer from a wheelchair or mobility aid to the *amusement ride seat*.

1002.5.4 Wheelchair Storage Space. Wheelchair storage *spaces* complying with 305 shall be provided in or adjacent to unload areas for each required *amusement ride seat* designed for transfer and shall not overlap any required means of egress or *accessible* route.

1002.6 Transfer Devices for Use with Amusement Rides. *Transfer devices* for use with *amusement rides* shall comply with 1002.6 when positioned for loading and unloading.

> **Advisory 1002.6 Transfer Devices for Use with Amusement Rides.** Transfer devices for use with amusement rides should permit individuals to make independent transfers to and from their wheelchairs or mobility devices. There are a variety of transfer devices available that could be adapted to provide access onto an amusement ride. Examples of devices that may provide for transfers include, but are not limited to, transfer systems, lifts, mechanized seats, and custom designed systems. Operators and designers have flexibility in developing designs that will facilitate individuals to transfer onto amusement rides. These systems or devices should be designed to be reliable and sturdy.
>
> Designs that limit the number of transfers required from a wheelchair or mobility device to the ride seat are encouraged. When using a transfer device to access an amusement ride, the least number of transfers and the shortest distance is most usable. Where possible, designers are encouraged to locate the transfer device seat no higher than 17 to 19 inches (430 to 485 mm) above the load and unload surface. Where greater distances are required for transfers, providing gripping surfaces, seat padding, and avoiding sharp objects in the path of transfer will facilitate the transfer. Where a series of transfers are required to reach the amusement ride seat, each vertical transfer should not exceed 8 inches (205 mm).

1002.6.1 Clear Floor or Ground Space. A clear floor or ground *space* complying with 305 shall be provided in the load and unload area adjacent to the *transfer device*.

1002.6.2 Transfer Height. The height of *transfer device* seats shall be 14 inches (355 mm) minimum and 24 inches (610 mm) maximum measured from the load and unload surface.

1002.6.3 Wheelchair Storage Space. Wheelchair storage *spaces* complying with 305 shall be provided in or adjacent to unload areas for each required *transfer device* and shall not overlap any required means of egress or *accessible* route.

1003 Recreational Boating Facilities

1003.1 General. Recreational boating *facilities* shall comply with 1003.

1003.2 Accessible Routes. *Accessible* routes serving recreational boating *facilities*, including *gangways* and floating piers, shall comply with Chapter 4 except as modified by the exceptions in 1003.2.

1003.2.1 Boat Slips. *Accessible* routes serving *boat slips* shall be permitted to use the exceptions in 1003.2.1.

EXCEPTIONS: 1. Where an existing *gangway* or series of *gangways* is replaced or *altered*, an increase in the length of the *gangway* shall not be required to comply with 1003.2 unless required by 202.4.
2. *Gangways* shall not be required to comply with the maximum rise specified in 405.6.
3. Where the total length of a *gangway* or series of *gangways* serving as part of a required *accessible* route is 80 feet (24 m) minimum, *gangways* shall not be required to comply with 405.2.
4. Where *facilities* contain fewer than 25 *boat slips* and the total length of the *gangway* or series of *gangways* serving as part of a required *accessible* route is 30 feet (9145 mm) minimum, *gangways* shall not be required to comply with 405.2.
5. Where *gangways* connect to *transition plates*, landings specified by 405.7 shall not be required.
6. Where *gangways* and *transition plates* connect and are required to have handrails, handrail extensions shall not be required. Where handrail extensions are provided on *gangways* or *transition plates*, the handrail extensions shall not be required to be parallel with the ground or floor surface.
7. The *cross slope* specified in 403.3 and 405.3 for *gangways*, *transition plates*, and floating piers that are part of *accessible* routes shall be measured in the static position.
8. Changes in level complying with 303.3 and 303.4 shall be permitted on the surfaces of *gangways* and *boat launch ramps*.

> **Advisory 1003.2.1 Boat Slips Exception 3.** The following example shows how exception 3 would be applied: A gangway is provided to a floating pier which is required to be on an accessible route. The vertical distance is 10 feet (3050 mm) between the elevation where the gangway departs the landside connection and the elevation of the pier surface at the lowest water level. Exception 3 permits the gangway to be 80 feet (24 m) long. Another design solution would be to have two 40 foot (12 m) plus continuous gangways joined together at a float, where the float (as the water level falls) will stop dropping at an elevation five feet below the landside connection. The length of transition plates would not be included in determining if the gangway(s) meet the requirements of the exception.

1003.2.2 Boarding Piers at Boat Launch Ramps. *Accessible* routes serving *boarding piers* at *boat launch ramps* shall be permitted to use the exceptions in 1003.2.2.
> EXCEPTIONS: 1. *Accessible* routes serving floating *boarding piers* shall be permitted to use Exceptions 1, 2, 5, 6, 7 and 8 in 1003.2.1.
> 2. Where the total length of the *gangway* or series of *gangways* serving as part of a required *accessible* route is 30 feet (9145 mm) minimum, *gangways* shall not be required to comply with 405.2.
> 3. Where the *accessible* route serving a floating *boarding pier* or skid pier is located within a *boat launch ramp*, the portion of the *accessible* route located within the *boat launch ramp* shall not be required to comply with 405.

1003.3 Clearances. Clearances at *boat slips* and on *boarding piers* at *boat launch ramps* shall comply with 1003.3.

> **Advisory 1003.3 Clearances.** Although the minimum width of the clear pier space is 60 inches (1525 mm), it is recommended that piers be wider than 60 inches (1525 mm) to improve the safety for persons with disabilities, particularly on floating piers.

1003.3.1 Boat Slip Clearance. *Boat slips* shall provide clear pier *space* 60 inches (1525 mm) wide minimum and at least as long as the *boat slips*. Each 10 feet (3050 mm) maximum of linear pier edge serving *boat slips* shall contain at least one continuous clear opening 60 inches (1525 mm) wide minimum.
> EXCEPTIONS: 1. Clear pier *space* shall be permitted to be 36 inches (915 mm) wide minimum for a length of 24 inches (610 mm) maximum, provided that multiple 36 inch (915 mm) wide segments are separated by segments that are 60 inches (1525 mm) wide minimum and 60 inches (1525 mm) long minimum.
> 2. Edge protection shall be permitted at the continuous clear openings, provided that it is 4 inches (100 mm) high maximum and 2 inches (51 mm) wide maximum.
> 3. In existing piers, clear pier *space* shall be permitted to be located perpendicular to the *boat slip* and shall extend the width of the *boat slip*, where the *facility* has at least one *boat slip* complying with 1003.3, and further compliance with 1003.3 would result in a reduction in the number of *boat slips* available or result in a reduction of the widths of existing slips.

TECHNICAL   CHAPTER 10: RECREATION FACILITIES

> **Advisory 1003.3.1 Boat Slip Clearance Exception 3.** Where the conditions in exception 3 are satisfied, existing facilities are only required to have one accessible boat slip with a pier clearance which runs the length of the slip. All other accessible slips are allowed to have the required pier clearance at the head of the slip. Under this exception, at piers with perpendicular boat slips, the width of most "finger piers" will remain unchanged. However, where mooring systems for floating piers are replaced as part of pier alteration projects, an opportunity may exist for increasing accessibility. Piers may be reconfigured to allow an increase in the number of wider finger piers, and serve as accessible boat slips.

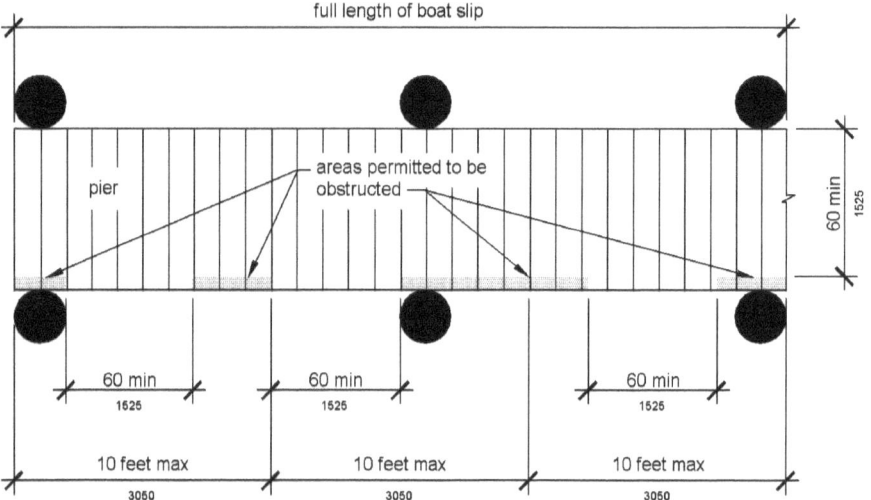

Figure 1003.3.1
Boat Slip Clearance

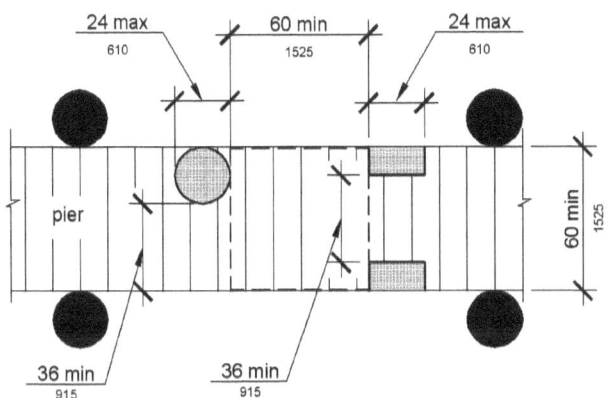

Figure 1003.3.1 (Exception 1)
Clear Pier Space Reduction at Boat Slips

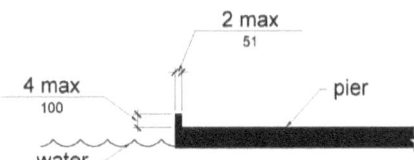

Figure 1003.3.1 (Exception 2)
Edge Protection at Boat Slips

1003.3.2 Boarding Pier Clearances. *Boarding piers* at *boat launch ramps* shall provide clear pier *space* 60 inches (1525 mm) wide minimum and shall extend the full length of the *boarding pier*. Every 10 feet (3050 mm) maximum of linear pier edge shall contain at least one continuous clear opening 60 inches (1525 mm) wide minimum.

    EXCEPTIONS: 1. The clear pier *space* shall be permitted to be 36 inches (915 mm) wide minimum for a length of 24 inches (610 mm) maximum provided that multiple 36 inch (915 mm) wide segments are separated by segments that are 60 inches (1525 mm) wide minimum and 60 inches (1525 mm) long minimum.

    2. Edge protection shall be permitted at the continuous clear openings provided that it is 4 inches (100 mm) high maximum and 2 inches (51 mm) wide maximum.

Advisory 1003.3.2 Boarding Pier Clearances. These requirements do not establish a minimum length for accessible boarding piers at boat launch ramps. The accessible boarding pier should have a length at least equal to that of other boarding piers provided at the facility. If no other boarding pier is provided, the pier would have a length equal to what would have been provided if no access requirements applied. The entire length of accessible boarding piers would be required to comply with the same technical provisions that apply to accessible boat slips. For example, at a launch ramp, if a 20 foot (6100 mm) long accessible boarding pier is provided, the entire 20 feet (6100 mm) must comply with the pier clearance requirements in 1003.3. Likewise, if a 60 foot (18 m) long accessible boarding pier is provided, the pier clearance requirements in 1003.3 would apply to the entire 60 feet (18 m).

The following example applies to a boat launch ramp boarding pier: A chain of floats is provided on a launch ramp to be used as a boarding pier which is required to be accessible by 1003.3.2. At high water, the entire chain is floating and a transition plate connects the first float to the surface of the launch ramp. As the water level decreases, segments of the chain end up resting on the launch ramp surface, matching the slope of the launch ramp.

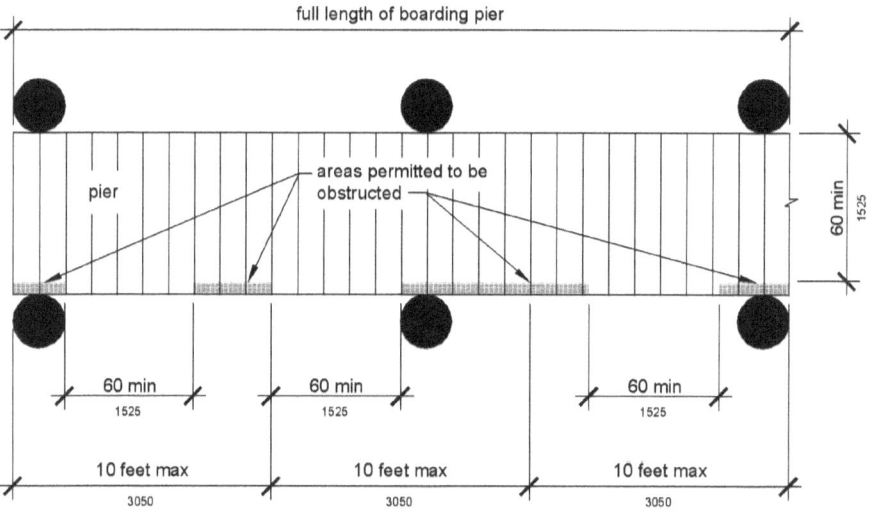

Figure 1003.3.2
Boarding Pier Clearance

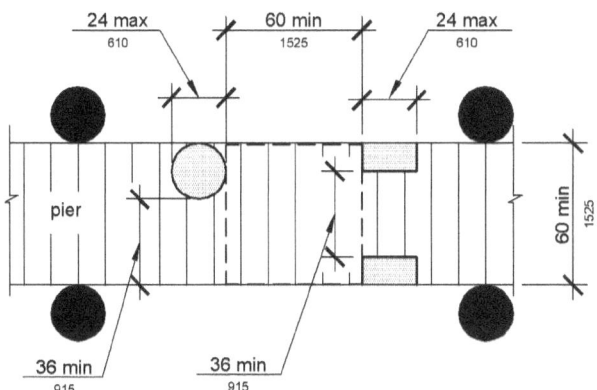

Figure 1003.3.2 (Exception 1)
Clear Pier Space Reduction at Boarding Piers

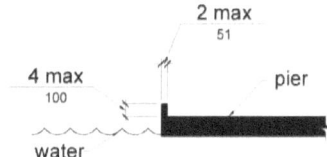

Figure 1003.3.2 (Exception 2)
Edge Protection at Boarding Piers

1004 Exercise Machines and Equipment

1004.1 Clear Floor Space. Exercise machines and equipment shall have a clear floor *space* complying with 305 positioned for transfer or for use by an individual seated in a wheelchair. Clear floor or ground *spaces* required at exercise machines and equipment shall be permitted to overlap.

> **Advisory 1004.1 Clear Floor Space.** One clear floor or ground space is permitted to be shared between two pieces of exercise equipment. To optimize space use, designers should carefully consider layout options such as connecting ends of the row and center aisle spaces. The position of the clear floor space may vary greatly depending on the use of the equipment or machine. For example, to provide access to a shoulder press machine, clear floor space next to the seat would be appropriate to allow for transfer. Clear floor space for a bench press machine designed for use by an individual seated in a wheelchair, however, will most likely be centered on the operating mechanisms.

1005 Fishing Piers and Platforms

1005.1 Accessible Routes. *Accessible* routes serving fishing piers and platforms, including *gangways* and floating piers, shall comply with Chapter 4.

> **EXCEPTIONS: 1.** *Accessible* routes serving floating fishing piers and platforms shall be permitted to use Exceptions 1, 2, 5, 6, 7 and 8 in 1003.2.1.
>
> **2.** Where the total length of the *gangway* or series of *gangways* serving as part of a required *accessible* route is 30 feet (9145 mm) minimum, *gangways* shall not be required to comply with 405.2.

1005.2 Railings. Where provided, railings, guards, or handrails shall comply with 1005.2.

1005.2.1 Height. At least 25 percent of the railings, guards, or handrails shall be 34 inches (865 mm) maximum above the ground or deck surface.

> **EXCEPTION:** Where a guard complying with sections 1003.2.12.1 and 1003.2.12.2 of the International Building Code (2000 edition) or sections 1012.2 and 1012.3 of the International Building Code (2003 edition) (incorporated by reference, see "Referenced Standards" in Chapter 1) is provided, the guard shall not be required to comply with 1005.2.1.

1005.2.1.1 Dispersion. Railings, guards, or handrails required to comply with 1005.2.1 shall be dispersed throughout the fishing pier or platform.

> **Advisory 1005.2.1.1 Dispersion.** Portions of the railings that are lowered to provide fishing opportunities for persons with disabilities must be located in a variety of locations on the fishing pier or platform to give people a variety of locations to fish. Different fishing locations may provide varying water depths, shade (at certain times of the day), vegetation, and proximity to the shoreline or bank.

1005.3 Edge Protection. Where railings, guards, or handrails complying with 1005.2 are provided, edge protection complying with 1005.3.1 or 1005.3.2 shall be provided.

> **Advisory 1005.3 Edge Protection.** Edge protection is required only where railings, guards, or handrails are provided on a fishing pier or platform. Edge protection will prevent wheelchairs or other mobility devices from slipping off the fishing pier or platform. Extending the deck of the fishing pier or platform 12 inches (305 mm) where the 34 inch (865 mm) high railing is provided is an alternative design, permitting individuals using wheelchairs or other mobility devices to pull into a clear space and move beyond the face of the railing. In such a design, curbs or barriers are not required.

1005.3.1 Curb or Barrier. Curbs or barriers shall extend 2 inches (51 mm) minimum above the surface of the fishing pier or platform.

1005.3.2 Extended Ground or Deck Surface. The ground or deck surface shall extend 12 inches (305 mm) minimum beyond the inside face of the railing. Toe clearance shall be provided and shall be 30 inches (760 mm) wide minimum and 9 inches (230 mm) minimum above the ground or deck surface beyond the railing.

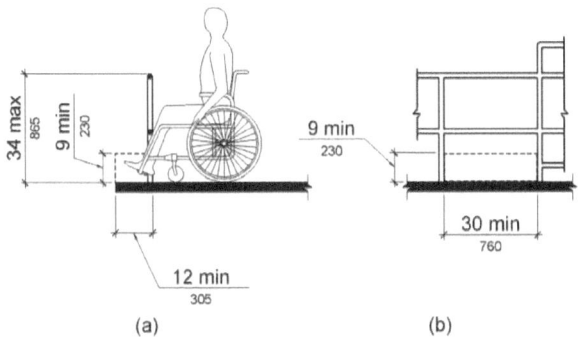

Figure 1005.3.2
Extended Ground or Deck Surface at Fishing Piers and Platforms

1005.4 Clear Floor or Ground Space. At each location where there are railings, guards, or handrails complying with 1005.2.1, a clear floor or ground *space* complying with 305 shall be provided. Where there are no railings, guards, or handrails, at least one clear floor or ground *space* complying with 305 shall be provided on the fishing pier or platform.

1005.5 Turning Space. At least one turning *space* complying with 304.3 shall be provided on fishing piers and platforms.

1006 Golf Facilities

1006.1 General. Golf *facilities* shall comply with 1006.

1006.2 Accessible Routes. *Accessible* routes serving *teeing grounds*, practice *teeing grounds*, putting greens, practice putting greens, teeing stations at driving ranges, course weather shelters, golf car rental areas, bag drop areas, and course toilet rooms shall comply with Chapter 4 and shall be 48 inches (1220 mm) wide minimum. Where handrails are provided, *accessible* routes shall be 60 inches (1525 mm) wide minimum.
   EXCEPTION: Handrails shall not be required on golf courses. Where handrails are provided on golf courses, the handrails shall not be required to comply with 505.

> **Advisory 1006.2 Accessible Routes.** The 48 inch (1220 mm) minimum width for the accessible route is necessary to ensure passage of a golf car on either the accessible route or the golf car passage. This is important where the accessible route is used to connect the golf car rental area, bag drop areas, practice putting greens, practice teeing grounds, course toilet rooms, and course weather shelters. These are areas outside the boundary of the golf course, but are areas where an individual using an adapted golf car may travel. A golf car passage may not be substituted for other accessible routes to be located outside the boundary of the course. For example, an accessible route connecting an accessible parking space to the entrance of a golf course clubhouse is not covered by this provision.
>
> Providing a golf car passage will permit a person that uses a golf car to practice driving a golf ball from the same position and stance used when playing the game. Additionally, the space required for a person using a golf car to enter and maneuver within the teeing stations required to be accessible should be considered.

1006.3 Golf Car Passages. *Golf car passages* shall comply with 1006.3.

1006.3.1 Clear Width. The clear width of *golf car passages* shall be 48 inches (1220 mm) minimum.

1006.3.2 Barriers. Where curbs or other constructed barriers prevent golf cars from entering a fairway, openings 60 inches (1525 mm) wide minimum shall be provided at intervals not to exceed 75 yards (69 m).

1006.4 Weather Shelters. A clear floor or ground *space* 60 inches (1525 mm) minimum by 96 inches (2440 mm) minimum shall be provided within weather shelters.

1007 Miniature Golf Facilities

1007.1 General. Miniature golf *facilities* shall comply with 1007.

1007.2 Accessible Routes. *Accessible* routes serving holes on miniature golf courses shall comply with Chapter 4. *Accessible* routes located on playing surfaces of miniature golf holes shall be permitted to use the exceptions in 1007.2.
   EXCEPTIONS: 1. Playing surfaces shall not be required to comply with 302.2.
   2. Where *accessible* routes intersect playing surfaces of holes, a 1 inch (25 mm) maximum curb shall be permitted for a width of 32 inches (815 mm) minimum.
   3. A slope not steeper than 1:4 for a 4 inch (100 mm) maximum rise shall be permitted.
   4. *Ramp* landing slopes specified by 405.7.1 shall be permitted to be 1:20 maximum.
   5. *Ramp* landing length specified by 405.7.3 shall be permitted to be 48 inches (1220 mm) long minimum.
   6. *Ramp* landing size specified by 405.7.4 shall be permitted to be 48 inches (1220 mm) minimum by 60 inches (1525 mm) minimum.
   7. Handrails shall not be required on holes. Where handrails are provided on holes, the handrails shall not be required to comply with 505.

1007.3 Miniature Golf Holes. Miniature golf holes shall comply with 1007.3.

1007.3.1 Start of Play. A clear floor or ground *space* 48 inches (1220 mm) minimum by 60 inches (1525 mm) minimum with slopes not steeper than 1:48 shall be provided at the start of play.

1007.3.2 Golf Club Reach Range Area. All areas within holes where golf balls rest shall be within 36 inches (915 mm) maximum of a clear floor or ground *space* 36 inches (915 mm) wide minimum and 48 inches (1220 mm) long minimum having a *running slope* not steeper than 1:20. The clear floor or ground *space* shall be served by an *accessible* route.

> **Advisory 1007.3.2 Golf Club Reach Range Area.** The golf club reach range applies to all holes required to be accessible. This includes accessible routes provided adjacent to or, where provided, on the playing surface of the hole.

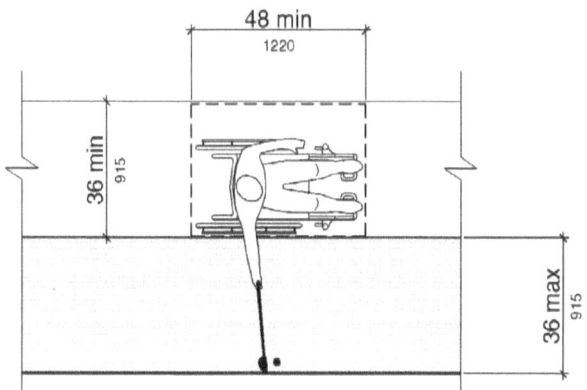

Note: Running Slope of Clear Floor or Ground Space Not Steeper Than 1:20

**Figure 1007.3.2
Golf Club Reach Range Area**

1008 Play Areas

1008.1 General. *Play areas* shall comply with 1008.

1008.2 Accessible Routes. *Accessible* routes serving *play areas* shall comply with Chapter 4 and 1008.2 and shall be permitted to use the exceptions in 1008.2.1 through 1008.2.3. Where *accessible* routes serve *ground level play components*, the vertical clearance shall be 80 inches high (2030 mm) minimum.

1008.2.1 Ground Level and Elevated Play Components. *Accessible* routes serving *ground level play components* and *elevated play components* shall be permitted to use the exceptions in 1008.2.1.

EXCEPTIONS: 1. Transfer systems complying with 1008.3 shall be permitted to connect *elevated play components* except where 20 or more *elevated play components* are provided no more than 25 percent of the *elevated play components* shall be permitted to be connected by transfer systems.
2. Where transfer systems are provided, an *elevated play component* shall be permitted to connect to another *elevated play component* as part of an *accessible* route.

1008.2.2 Soft Contained Play Structures. *Accessible* routes serving *soft contained play structures* shall be permitted to use the exception in 1008.2.2.
EXCEPTION: Transfer systems complying with 1008.3 shall be permitted to be used as part of an *accessible* route.

1008.2.3 Water Play Components. *Accessible* routes serving water *play components* shall be permitted to use the exceptions in 1008.2.3.
EXCEPTIONS: 1. Where the surface of the *accessible* route, clear floor or ground *spaces*, or turning *spaces* serving water *play components* is submerged, compliance with 302, 403.3, 405.2, 405.3, and 1008.2.6 shall not be required.
2. Transfer systems complying with 1008.3 shall be permitted to connect *elevated play components* in water.

> **Advisory 1008.2.3 Water Play Components.** Personal wheelchairs and mobility devices may not be appropriate for submerging in water when using play components in water. Some may have batteries, motors, and electrical systems that when submerged in water may cause damage to the personal mobility device or wheelchair or may contaminate the water. Providing an aquatic wheelchair made of non-corrosive materials and designed for access into the water will protect the water from contamination and avoid damage to personal wheelchairs.

1008.2.4 Clear Width. *Accessible* routes connecting *play components* shall provide a clear width complying with 1008.2.4.

1008.2.4.1 Ground Level. At ground level, the clear width of *accessible* routes shall be 60 inches (1525 mm) minimum.
EXCEPTIONS: 1. In *play areas* less than 1000 square feet (93 m^2), the clear width of *accessible* routes shall be permitted to be 44 inches (1120 mm) minimum, if at least one turning *space* complying with 304.3 is provided where the restricted *accessible* route exceeds 30 feet (9145 mm) in length.
2. The clear width of *accessible* routes shall be permitted to be 36 inches (915 mm) minimum for a distance of 60 inches (1525 mm) maximum provided that multiple reduced width segments are separated by segments that are 60 inches (1525 mm) wide minimum and 60 inches (1525 mm) long minimum.

1008.2.4.2 Elevated. The clear width of *accessible* routes connecting *elevated play components* shall be 36 inches (915 mm) minimum.

EXCEPTIONS: 1. The clear width of *accessible* routes connecting *elevated play components* shall be permitted to be reduced to 32 inches (815 mm) minimum for a distance of 24 inches (610 mm) maximum provided that reduced width segments are separated by segments that are 48 inches (1220 mm) long minimum and 36 inches (915 mm) wide minimum.
2. The clear width of transfer systems connecting *elevated play components* shall be permitted to be 24 inches (610 mm) minimum.

1008.2.5 Ramps. Within *play areas*, *ramps* connecting *ground level play components* and *ramps* connecting *elevated play components* shall comply with 1008.2.5.

1008.2.5.1 Ground Level. *Ramp* runs connecting *ground level play components* shall have a *running slope* not steeper than 1:16.

1008.2.5.2 Elevated. The rise for any *ramp* run connecting *elevated play components* shall be 12 inches (305 mm) maximum.

1008.2.5.3 Handrails. Where required on *ramps* serving *play components*, the handrails shall comply with 505 except as modified by 1008.2.5.3.
EXCEPTIONS: 1. Handrails shall not be required on *ramps* located within ground level *use zones*.
2. Handrail extensions shall not be required.

1008.2.5.3.1 Handrail Gripping Surfaces. Handrail gripping surfaces with a circular cross section shall have an outside diameter of 0.95 inch (24 mm) minimum and 1.55 inches (39 mm) maximum. Where the shape of the gripping surface is non-circular, the handrail shall provide an equivalent gripping surface.

1008.2.5.3.2 Handrail Height. The top of handrail gripping surfaces shall be 20 inches (510 mm) minimum and 28 inches (710 mm) maximum above the *ramp* surface.

1008.2.6 Ground Surfaces. Ground surfaces on *accessible* routes, clear floor or ground *spaces*, and turning *spaces* shall comply with 1008.2.6.

> **Advisory 1008.2.6 Ground Surfaces.** Ground surfaces must be inspected and maintained regularly to ensure continued compliance with the ASTM F 1951 standard. The type of surface material selected and play area use levels will determine the frequency of inspection and maintenance activities.

1008.2.6.1 Accessibility. Ground surfaces shall comply with ASTM F 1951 (incorporated by reference, see "Referenced Standards" in Chapter 1). Ground surfaces shall be inspected and maintained regularly and frequently to ensure continued compliance with ASTM F 1951.

1008.2.6.2 Use Zones. Ground surfaces located within *use zones* shall comply with ASTM F 1292 (1999 edition or 2004 edition) (incorporated by reference, see "Referenced Standards" in Chapter 1).

1008.3 Transfer Systems. Where transfer systems are provided to connect to *elevated play components*, transfer systems shall comply with 1008.3.

> **Advisory 1008.3 Transfer Systems.** Where transfer systems are provided, consideration should be given to the distance between the transfer system and the elevated play components. Moving between a transfer platform and a series of transfer steps requires extensive exertion for some children. Designers should minimize the distance between the points where a child transfers from a wheelchair or mobility device and where the elevated play components are located. Where elevated play components are used to connect to another elevated play component instead of an accessible route, careful consideration should be used in the selection of the play components used for this purpose.

1008.3.1 Transfer Platforms. Transfer platforms shall be provided where transfer is intended from wheelchairs or other mobility aids. Transfer platforms shall comply with 1008.3.1.

1008.3.1.1 Size. Transfer platforms shall have level surfaces 14 inches (355 mm) deep minimum and 24 inches (610 mm) wide minimum.

1008.3.1.2 Height. The height of transfer platforms shall be 11 inches (280 mm) minimum and 18 inches (455 mm) maximum measured to the top of the surface from the ground or floor surface.

1008.3.1.3 Transfer Space. A transfer *space* complying with 305.2 and 305.3 shall be provided adjacent to the transfer platform. The 48 inch (1220 mm) long minimum dimension of the transfer *space* shall be centered on and parallel to the 24 inch (610 mm) long minimum side of the transfer platform. The side of the transfer platform serving the transfer *space* shall be unobstructed.

1008.3.1.4 Transfer Supports. At least one means of support for transferring shall be provided.

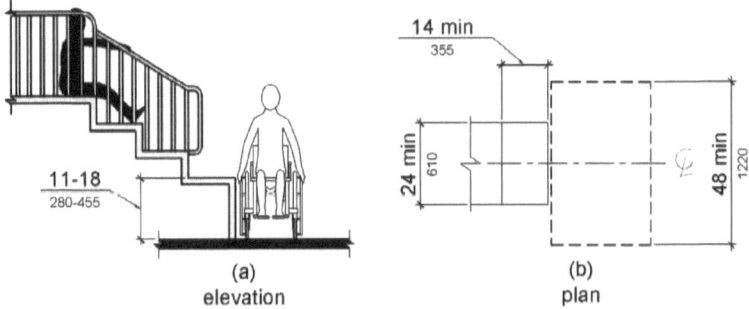

**Figure 1008.3.1
Transfer Platforms**

CHAPTER 10: RECREATION FACILITIES — TECHNICAL

1008.3.2 Transfer Steps. Transfer steps shall be provided where movement is intended from transfer platforms to levels with *elevated play components* required to be on *accessible* routes. Transfer steps shall comply with 1008.3.2.

1008.3.2.1 Size. Transfer steps shall have level surfaces 14 inches (355 mm) deep minimum and 24 inches (610 mm) wide minimum.

1008.3.2.2 Height. Each transfer step shall be 8 inches (205 mm) high maximum.

1008.3.2.3 Transfer Supports. At least one means of support for transferring shall be provided.

> **Advisory 1008.3.2.3 Transfer Supports.** Transfer supports are required on transfer platforms and transfer steps to assist children when transferring. Some examples of supports include a rope loop, a loop type handle, a slot in the edge of a flat horizontal or vertical member, poles or bars, or D rings on the corner posts.

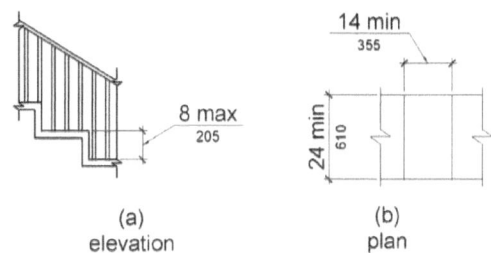

Figure 1008.3.2
Transfer Steps

1008.4 Play Components. *Ground level play components* on *accessible* routes and *elevated play components* connected by *ramps* shall comply with 1008.4.

1008.4.1 Turning Space. At least one turning *space* complying with 304 shall be provided on the same level as *play components*. Where swings are provided, the turning *space* shall be located immediately adjacent to the swing.

1008.4.2 Clear Floor or Ground Space. Clear floor or ground *space* complying with 305.2 and 305.3 shall be provided at *play components*.

Advisory 1008.4.2 Clear Floor or Ground Space. Clear floor or ground spaces, turning spaces, and accessible routes are permitted to overlap within play areas. A specific location has not been designated for the clear floor or ground spaces or turning spaces, except swings, because each play component may require that the spaces be placed in a unique location. Where play components include a seat or entry point, designs that provide for an unobstructed transfer from a wheelchair or other mobility device are recommended. This will enhance the ability of children with disabilities to independently use the play component.

When designing play components with manipulative or interactive features, consider appropriate reach ranges for children seated in wheelchairs. The following table provides guidance on reach ranges for children seated in wheelchairs. These dimensions apply to either forward or side reaches. The reach ranges are appropriate for use with those play components that children seated in wheelchairs may access and reach. Where transfer systems provide access to elevated play components, the reach ranges are not appropriate.

Children's Reach Ranges			
Forward or Side Reach	Ages 3 and 4	Ages 5 through 8	Ages 9 through 12
High (maximum)	36 in (915 mm)	40 in (1015 mm)	44 in (1120 mm)
Low (minimum)	20 in (510 mm)	18 in (455 mm)	16 in (405 mm)

1008.4.3 Play Tables. Where play tables are provided, knee clearance 24 inches (610 mm) high minimum, 17 inches deep (430 mm) minimum, and 30 inches (760 mm) wide minimum shall be provided. The tops of rims, curbs, or other obstructions shall be 31 inches (785 mm) high maximum.

EXCEPTION: Play tables designed and constructed primarily for children 5 years and younger shall not be required to provide knee clearance where the clear floor or ground *space* required by 1008.4.2 is arranged for a parallel approach.

1008.4.4 Entry Points and Seats. Where *play components* require transfer to entry points or seats, the entry points or seats shall be 11 inches (280 mm) minimum and 24 inches (610 mm) maximum from the clear floor or ground *space*.

EXCEPTION: Entry points of slides shall not be required to comply with 1008.4.4.

1008.4.5 Transfer Supports. Where *play components* require transfer to entry points or seats, at least one means of support for transferring shall be provided.

1009 Swimming Pools, Wading Pools, and Spas

1009.1 General. Where provided, pool lifts, sloped entries, transfer walls, transfer systems, and pool stairs shall comply with 1009.

1009.2 Pool Lifts. Pool lifts shall comply with 1009.2.

> **Advisory 1009.2 Pool Lifts.** There are a variety of seats available on pool lifts ranging from sling seats to those that are preformed or molded. Pool lift seats with backs will enable a larger population of persons with disabilities to use the lift. Pool lift seats that consist of materials that resist corrosion and provide a firm base to transfer will be usable by a wider range of people with disabilities. Additional options such as armrests, head rests, seat belts, and leg support will enhance accessibility and better accommodate people with a wide range of disabilities.

1009.2.1 Pool Lift Location. Pool lifts shall be located where the water level does not exceed 48 inches (1220 mm).

    EXCEPTIONS: 1. Where the entire pool depth is greater than 48 inches (1220 mm), compliance with 1009.2.1 shall not be required.

    2. Where multiple pool lift locations are provided, no more than one pool lift shall be required to be located in an area where the water level is 48 inches (1220 mm) maximum.

1009.2.2 Seat Location. In the raised position, the centerline of the seat shall be located over the deck and 16 inches (405 mm) minimum from the edge of the pool. The deck surface between the centerline of the seat and the pool edge shall have a slope not steeper than 1:48.

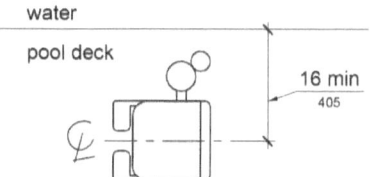

Figure 1009.2.2
Pool Lift Seat Location

1009.2.3 Clear Deck Space. On the side of the seat opposite the water, a clear deck *space* shall be provided parallel with the seat. The *space* shall be 36 inches (915 mm) wide minimum and shall extend forward 48 inches (1220 mm) minimum from a line located 12 inches (305 mm) behind the rear edge of the seat. The clear deck *space* shall have a slope not steeper than 1:48.

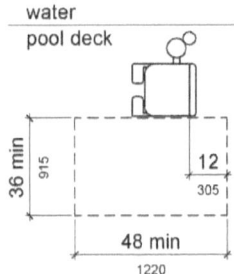

Figure 1009.2.3
Clear Deck Space at Pool Lifts

1009.2.4 Seat Height. The height of the lift seat shall be designed to allow a stop at 16 inches (405 mm) minimum to 19 inches (485 mm) maximum measured from the deck to the top of the seat surface when in the raised (load) position.

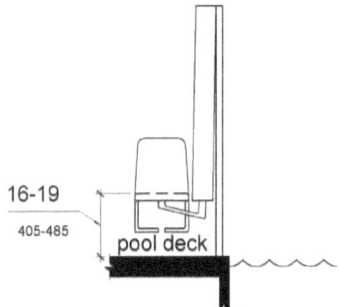

Figure 1009.2.4
Pool Lift Seat Height

1009.2.5 Seat Width. The seat shall be 16 inches (405 mm) wide minimum.

1009.2.6 Footrests and Armrests. Footrests shall be provided and shall move with the seat. If provided, the armrest positioned opposite the water shall be removable or shall fold clear of the seat when the seat is in the raised (load) position.
  EXCEPTION: Footrests shall not be required on pool lifts provided in spas.

1009.2.7 Operation. The lift shall be capable of unassisted operation from both the deck and water levels. Controls and operating mechanisms shall be unobstructed when the lift is in use and shall comply with 309.4.

> **Advisory 1009.2.7 Operation.** Pool lifts must be capable of unassisted operation from both the deck and water levels. This will permit a person to call the pool lift when the pool lift is in the opposite position. It is extremely important for a person who is swimming alone to be able to call the pool lift when it is in the up position so he or she will not be stranded in the water for extended periods of time awaiting assistance. The requirement for a pool lift to be independently operable does not preclude assistance from being provided.

1009.2.8 Submerged Depth. The lift shall be designed so that the seat will submerge to a water depth of 18 inches (455 mm) minimum below the stationary water level.

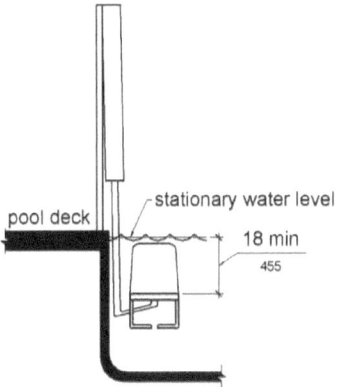

Figure 1009.2.8
Pool Lift Submerged Depth

1009.2.9 Lifting Capacity. Single person pool lifts shall have a weight capacity of 300 pounds. (136 kg) minimum and be capable of sustaining a static load of at least one and a half times the rated load.

> **Advisory 1009.2.9 Lifting Capacity.** Single person pool lifts must be capable of supporting a minimum weight of 300 pounds (136 kg) and sustaining a static load of at least one and a half times the rated load. Pool lifts should be provided that meet the needs of the population they serve. Providing a pool lift with a weight capacity greater than 300 pounds (136 kg) may be advisable.

1009.3 Sloped Entries. Sloped entries shall comply with 1009.3.

> **Advisory 1009.3 Sloped Entries.** Personal wheelchairs and mobility devices may not be appropriate for submerging in water. Some may have batteries, motors, and electrical systems that when submerged in water may cause damage to the personal mobility device or wheelchair or may contaminate the pool water. Providing an aquatic wheelchair made of non-corrosive materials and designed for access into the water will protect the water from contamination and avoid damage to personal wheelchairs or other mobility aids.

1009.3.1 Sloped Entries. Sloped entries shall comply with Chapter 4 except as modified in 1109.3.1 through 1109.3.3.

   EXCEPTION: Where sloped entries are provided, the surfaces shall not be required to be slip resistant.

1009.3.2 Submerged Depth. Sloped entries shall extend to a depth of 24 inches (610 mm) minimum and 30 inches (760 mm) maximum below the stationary water level. Where landings are required by 405.7, at least one landing shall be located 24 inches (610 mm) minimum and 30 inches (760 mm) maximum below the stationary water level.

   EXCEPTION: In wading pools, the sloped entry and landings, if provided, shall extend to the deepest part of the wading pool.

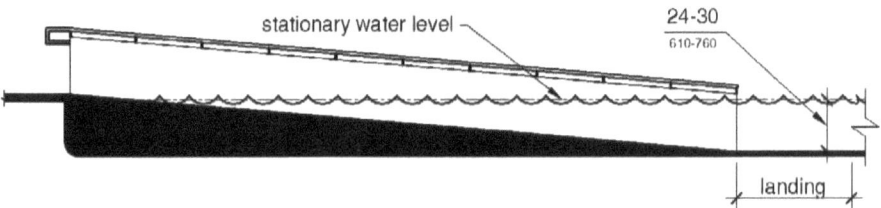

Figure 1009.3.2
Sloped Entry Submerged Depth

1009.3.3 Handrails. At least two handrails complying with 505 shall be provided on the sloped entry. The clear width between required handrails shall be 33 inches (840 mm) minimum and 38 inches (965 mm) maximum.

   EXCEPTIONS: 1. Handrail extensions specified by 505.10.1 shall not be required at the bottom landing serving a sloped entry.
   2. Where a sloped entry is provided for wave action pools, leisure rivers, sand bottom pools, and other pools where user access is limited to one area, the handrails shall not be required to comply with the clear width requirements of 1009.3.3.
   3. Sloped entries in wading pools shall not be required to provide handrails complying with 1009.3.3. If provided, handrails on sloped entries in wading pools shall not be required to comply with 505.

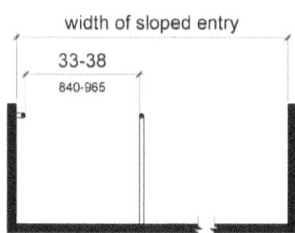

Figure 1009.3.3
Handrails for Sloped Entry

1009.4 Transfer Walls. Transfer walls shall comply with 1009.4.

1009.4.1 Clear Deck Space. A clear deck *space* of 60 inches (1525 mm) minimum by 60 inches (1525 mm) minimum with a slope not steeper than 1:48 shall be provided at the base of the transfer wall. Where one grab bar is provided, the clear deck *space* shall be centered on the grab bar. Where two grab bars are provided, the clear deck *space* shall be centered on the clearance between the grab bars.

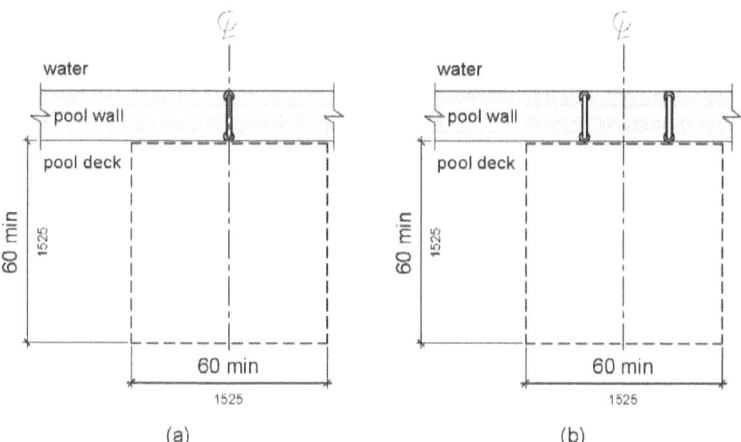

Figure 1009.4.1
Clear Deck Space at Transfer Walls

1009.4.2 Height. The height of the transfer wall shall be 16 inches (405 mm) minimum and 19 inches (485 mm) maximum measured from the deck.

Figure 1009.4.2
Transfer Wall Height

1009.4.3 Wall Depth and Length. The depth of the transfer wall shall be 12 inches (305 mm) minimum and 16 inches (405 mm) maximum. The length of the transfer wall shall be 60 inches (1525 mm) minimum and shall be centered on the clear deck *space*.

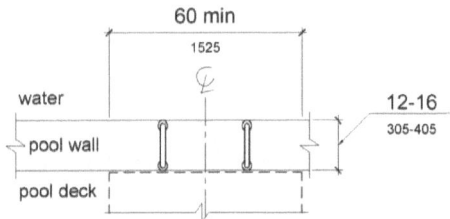

Figure 1009.4.3
Depth and Length of Transfer Walls

1009.4.4 Surface. Surfaces of transfer walls shall not be sharp and shall have rounded edges.

1009.4.5 Grab Bars. At least one grab bar complying with 609 shall be provided on the transfer wall. Grab bars shall be perpendicular to the pool wall and shall extend the full depth of the transfer wall. The top of the gripping surface shall be 4 inches (100 mm) minimum and 6 inches (150 mm) maximum above transfer walls. Where one grab bar is provided, clearance shall be 24 inches (610 mm) minimum on both sides of the grab bar. Where two grab bars are provided, clearance between grab bars shall be 24 inches (610 mm) minimum.

    EXCEPTION: Grab bars on transfer walls shall not be required to comply with 609.4.

CHAPTER 10: RECREATION FACILITIES   TECHNICAL

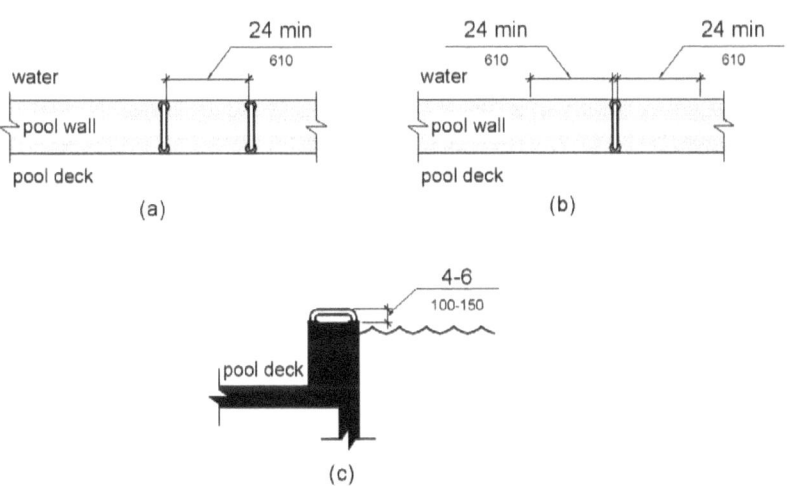

Figure 1009.4.5
Grab Bars for Transfer Walls

1009.5 Transfer Systems. Transfer systems shall comply with 1009.5.

1009.5.1 Transfer Platform. A transfer platform shall be provided at the head of each transfer system. Transfer platforms shall provide 19 inches (485 mm) minimum clear depth and 24 inches (610 mm) minimum clear width.

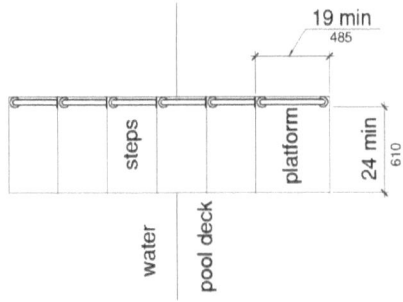

Figure 1009.5.1
Size of Transfer Platform

1009.5.2 Transfer Space. A transfer *space* of 60 inches (1525 mm) minimum by 60 inches (1525 mm) minimum with a slope not steeper than 1:48 shall be provided at the base of the transfer platform surface and shall be centered along a 24 inch (610 mm) minimum side of the transfer platform. The side of the transfer platform serving the transfer *space* shall be unobstructed.

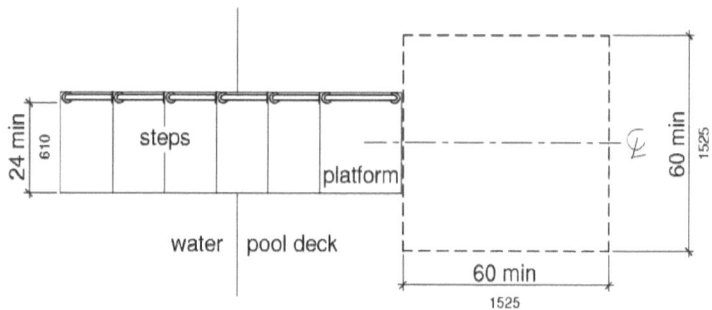

Figure 1009.5.2
Clear Deck Space at Transfer Platform

1009.5.3 Height. The height of the transfer platform shall comply with 1009.4.2.

1009.5.4 Transfer Steps. Transfer step height shall be 8 inches (205 mm) maximum. The surface of the bottom tread shall extend to a water depth of 18 inches (455 mm) minimum below the stationary water level.

Advisory 1009.5.4 Transfer Steps. Where possible, the height of the transfer step should be minimized to decrease the distance an individual is required to lift up or move down to reach the next step to gain access.

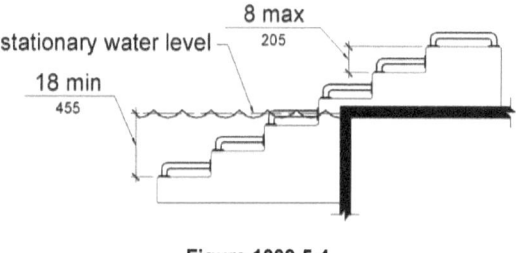

Figure 1009.5.4
Transfer Steps

1009.5.5 Surface. The surface of the transfer system shall not be sharp and shall have rounded edges.

1009.5.6 Size. Each transfer step shall have a tread clear depth of 14 inches (355 mm) minimum and 17 inches (430 mm) maximum and shall have a tread clear width of 24 inches (610 mm) minimum.

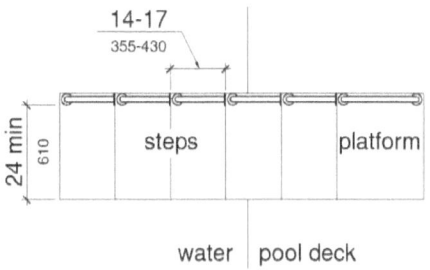

Figure 1009.5.6
Size of Transfer Steps

1009.5.7 Grab Bars. At least one grab bar on each transfer step and the transfer platform or a continuous grab bar serving each transfer step and the transfer platform shall be provided. Where a grab bar is provided on each step, the tops of gripping surfaces shall be 4 inches (100 mm) minimum and 6 inches (150 mm) maximum above each step and transfer platform. Where a continuous grab bar is provided, the top of the gripping surface shall be 4 inches (100 mm) minimum and 6 inches (150 mm) maximum above the step nosing and transfer platform. Grab bars shall comply with 609 and be located on at least one side of the transfer system. The grab bar located at the transfer platform shall not obstruct transfer.

EXCEPTION: Grab bars on transfer systems shall not be required to comply with 609.4.

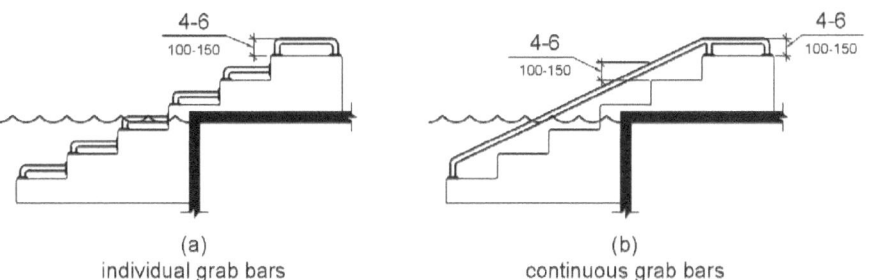

Figure 1009.5.7
Grab Bars

1009.6 Pool Stairs. Pool stairs shall comply with 1009.6.

1009.6.1 Pool Stairs. Pool stairs shall comply with 504.
EXCEPTION: Pool step riser heights shall not be required to be 4 inches (100 mm) high minimum and 7 inches (180 mm) high maximum provided that riser heights are uniform.

1009.6.2 Handrails. The width between handrails shall be 20 inches (510 mm) minimum and 24 inches (610 mm) maximum. Handrail extensions required by 505.10.3 shall not be required on pool stairs.

1010 Shooting Facilities with Firing Positions

1010.1 Turning Space. A circular turning *space* 60 inches (1525 mm) diameter minimum with slopes not steeper than 1:48 shall be provided at shooting facilities with firing positions.

INDEX TO THE 2010 STANDARDS

**36 CFR part 1191,
appendices B and D**
Application and Scoping
1

1991 Standards
Application and Scoping
8-11, 22, 26-28

2004 ADAAG
Application and Scoping
1, 14, 30

2010 Standards
Application and Scoping
1, 8, 10-14, 22, 26-30

A

Accessibility Standards
Application and Scoping
9, 22, 26

Accessible
Application and Scoping
44

Accessible Bathing Facility
Application and Scoping
28

**Accessible Entrances
(see Entrances)**

Accessible Features
Application and Scoping
8, 13, 24, 92

Accessible Rooms
Application and Scoping
28, 55, 88

Accessible Route(s)
Application and Scoping
8, 9, 11, 12, 21, 24, 29, 44, 50, 52-64, 66-69, 80, 81, 86, 92, 93, 98, 100-102

Technical
108, 113, 117, 118, 119, 127,131, 133,150-152, 162, 203, 212, 213, 215, 224, 226-229, 234-239, 241

Addition(s)
Application and Scoping
37, 44, 50, 57, 93, 99

Airport Passenger Terminal
Application and Scoping
21, 25, 56

Air Traffic Control Towers
Application and Scoping
56

**Aisle Seats (see Designated
Aisle Seats)**

Alteration(s)
Application and Scoping
6-10, 13, 14, 19, 21-28, 37, 44, 48, 50- 52, 55, 57, 58, 61, 68, 69, 81-83, 90, 93, 95

Technical
19, 131, 156, 212, 221

Amusement Ride(s)
Application and Scoping
45, 48, 59, 60, 64, 74, 94, 95

Technical
224-227

Index and List of Figures

Animal Containment Areas
Application and Scoping
54

Technical
104, 105

ANSI/BHMA
Application and Scoping
40

Technical
126, 127, 143, 145

Apartments
Application and Scoping
12, 29, 46

Area(s) of Sport Activity
Application and Scoping
45, 54, 64, 79, 80

Technical
104, 105

ASME
Application and Scoping
40, 41

Technical
133, 143, 145, 147, 218

Assembly Area(s)
Application and Scoping
12, 29, 45, 48, 57, 58, 68, 71, 73, 76-80

Technical
127, 154, 156, 197

Assistive Listening System(s)
Application and Scoping
45, 73, 76, 77

Technical
194, 197, 198

ASTM
Application and Scoping
41, 42, 48

Technical
239

ATMs (see Automatic Teller Machines)

Automatic Fare Machines (see Fare Machines)

Automatic Teller Machine(s)
Application and Scoping
77

Technical
198, 199, 201

B

Barrier Removal
Application and Scoping
37

Bathing Facilities
Application and Scoping
11, 29, 69, 70, 73

Technical
172, 177, 181, 210, 211, 213

Bathing Rooms
Application and Scoping
69, 70, 73

Technical
160, 210

Bathtubs
Application and Scoping
70

Technical
171-174, 182

Bathtub Seats (see Seats, Bathtub and Shower)

Beds
Application and Scoping
11, 28, 29, 84, 85, 90

Technical
210, 211

Bench(es) (does not include Judges' Benches)
Technical
185, 206, 211, 219, 220

Boarding Pier(s)
Application and Scoping
45, 60, 97

Technical
229, 231

Boat Launch Ramps
Application and Scoping
45, 60, 97

Technical
228, 229, 231

Boat Slip(s)
Application and Scoping
45, 60, 96

Technical
228, 229

Boating Facilities
Application and Scoping
60, 64, 95

Technical
228

Bowling Lanes
Application and Scoping
60, 79

Boxing or Wrestling Rings
Application and Scoping
54

Bus Shelters
Application and Scoping
76

Technical
215

Bus Signs
Technical
216

Bus Stop(s)
Application and Scoping
67, 68

Technical
214, 215, 216

C

Cafeterias (see Restaurants and Cafeterias)

Carpet
Technical
104

Cells
Application and Scoping
13, 14, 53, 56, 68, 89, 90, 91

Technical
163, 211, 212

Cells with Mobility Features
Application and Scoping
13, 14, 56, 90, 211

Change Machines
Application and Scoping
88

Changes in Level
Technical
105-107, 117, 124, 127-129, 151-153, 202, 228

Index and List of Figures

Check-out Aisles
Application and Scoping
73, 74, 87, 88

Technical
220, 221

Children
Application and Scoping
37, 42, 45, 47, 74, 95, 98-100

Technical
113, 154, 159, 160, 161, 165, 167-169, 171, 182, 206, 219, 240-242

Clear Floor or Ground Space
Application and Scoping
11, 28, 57, 84, 90

Technical
107-110 114-116, 127, 134, 138, 145, 147, 159, 160, 162, 170, 182, 185, 190, 194, 198, 201, 206, 208-212, 215, 218, 219, 221, 222, 227, 228, 233, 235-239, 241, 242

Clearances
Technical
117, 160-162, 174, 207, 225, 229, 231

Clearances, Maneuvering
Technical
120, 123, 124, 127,129

Clocks
Application and Scoping
76

Technical
218

Clothes Dryers
Application and Scoping
70, 71

Technical
115, 185

Coat Hooks
Application and Scoping
70, 81

Technical
161, 162, 168, 206

Commercial Facility
Application and Scoping
19, 21, 22

Commercial Facilities Located in a Private Residence
Application and Scoping
19

Common Ownership
Application and Scoping
20, 25

Common Use
Application and Scoping
11, 12, 29, 45, 47, 53, 56, 59, 71

Technical
117, 128, 129, 160, 163, 171, 172, 177, 201

Common Use Area
Application and Scoping
12, 29, 53, 56, 71

Common Use Circulation Paths
Application and Scoping
59

Technical
117, 128, 129

Common Site
Application and Scoping
20, 25, 28, 92, 93

Communication Systems (see Two-Way Communication Systems)

Commuter Rail (see also Rail Platforms)
Application and Scoping
46, 76

Index and List of Figures

Commuter Rail (cont'd)
Technical
217

Companion Seats
Application and Scoping
12, 29, 30, 78-80

Technical
202, 206, 226

Compliance Date
Application and Scoping
9, 10, 26, 27

Construction Sites
Application and Scoping
53

Conventions
Application and Scoping
37

Correctional Facilities (see Detention and Correctional Facilities)

Counters (see also Dining Surfaces and Work Surfaces)
Application and Scoping
59, 89, 91

Technical
207, 208, 219, 222

Counters, Sales and Service
Application and Scoping
86-88

Technical
220, 221

Court Sports
Application and Scoping
60

Courtroom
Application and Scoping
45, 54, 57, 63, 76, 89

Courtoom (con'td)
Technical
217

Cross Slope
Application and Scoping
45, 47

Technical
117, 128, 228

Curb Ramps
Application and Scoping
7, 13, 23, 45

Technical
117, 131-133

D

Definitions
Application and Scoping
44

Depot
Application and Scoping
21, 25, 56

Depositories
Application and Scoping
88

Designated Aisle Seats
Application and Scoping
78, 80

Technical
202, 206

Detectable Warnings
Application and Scoping
46

Technical
196, 197, 216

Index and List of Figures

Detention and Correctional Facilities
Application and Scoping
13, 14, 47, 48, 53, 56, 62, 64, 68, 71, 76, 89

Technical
163

Dimensions (General)
Application and Scoping
37, 38

Dining Surfaces
Application and Scoping
37, 86, 87

Technical
219

Disperse, Dispersion
Application and Scoping
12, 13, 29, 30, 63, 66, 67, 79, 80, 85-88, 94, 96, 99

Technical
234

Disproportionate, Disproportionality
Application and Scoping
7-9, 22-25, 51

Diving Boards and Diving Platforms
Application and Scoping
54

Door Swing
Technical
107, 127, 160, 190, 206

Doors, Doorways, and Gates (General)
Application and Scoping
8, 19, 24, 40, 46, 61-63, 70, 72, 83

Technical
117, 119, 120, 123-127, 129

Dressing, Fitting, and Locker Rooms
Application and Scoping
81

Dressing, Fitting, and Locker Rooms (cont'd)
Technical
206

Drinking Fountains
Application and Scoping
7-9, 22-24, 37, 51, 68, 69

Technical
159, 160

Dryers (see Washing Machines and Clothes Dryers)

E

Edge Protection
Technical
129, 130, 221, 229, 231, 234

Elevators
Application and Scoping
8, 23, 40, 45, 49, 63, 73

Technical
117, 133-138, 140-146, 189, 191

Elevator Exemption
Application and Scoping
9, 10, 20, 21, 25

Existing Elevator
Application and Scoping
63, 73

Technical
134, 135, 137, 138, 140, 141, 143

Employee Work Area
Application and Scoping
23, 46, 49, 54, 59, 71

Technical
117, 128, 129

Index and List of Figures

Entrances
Application and Scoping
7-9, 21-24, 46, 47, 48, 52, 55, 57, 61, 62, 66-69, 73, 76, 83, 98

Technical
152, 201, 214, 217

Public Entrance
Application and Scoping
47, 61

Restricted Entrance
Application and Scoping
47, 62

Service Entrance
Application and Scoping
47, 62

Equivalent Access
Application and Scoping
9, 10, 67, 79, 80, 94

Equivalent Facilitation
Application and Scoping
37

Exercise Machine
Application and Scoping
55, 60, 97

Technical
233

Existing Elevator (see Elevators)

Existing Buildings and Facilities
Application and Scoping
22, 37, 50, 57, 63, 71, 93

Technical
180

Exits (see Means of Egress)

F

Facilities with Residential Units and Transient Lodging Units
Application and Scoping
28

Fare Machines
Application and Scoping
77

Technical
198, 199, 201

Figures
Application and Scoping
11, 28, 38

Fire Alarm Systems
Application and Scoping
43, 71

Technical
186, 213

Fishing Piers and Platforms
Application and Scoping
60, 64, 97

Technical
234, 235

Fixed Guideway Stations
Application and Scoping
76

Floor or Ground Surfaces
Scoping
98, 99

Technical
104-107, 117, 124, 128, 130, 151, 152, 202, 224, 225, 239

Food Service Lines
Application and Scoping
87, 88

Index and List of Figures

Food Service Lines (cont'd)
Technical
222

Forward Reach (see Reach Ranges)

G

Gates (see Doors, Doorways, and Gates)

General Exceptions
Application and Scoping
53

Golf Car Passage
Application and Scoping
46, 60

Technical
236

Golf Facilities
Application and Scoping
60, 97

Technical
235

Grab Bars
Application and Scoping
8, 24

Technical
162-164, 167-169, 172-174, 177, 178, 180-182, 247, 248, 251

Gratings (see Openings)

Ground Level Play Components (see Play Areas)

Ground Surfaces (see Floor or Ground Surfaces)

Group Homes
Application and Scoping
11, 28

Guest Rooms (see Transient Lodging)

H

Halfway Houses
Application and Scoping
11, 28

Handrails
Application and Scoping
68

Technical
111, 112, 119, 128-130, 153-157, 224, 228, 234-236, 239, 246, 252

High Speed Rail
Application and Scoping
76

Historic Buildings and Facilities
Application and Scoping
7, 21, 26, 44, 47, 52, 55, 57, 61, 69

National Historic Preservation Act
Application and Scoping
26

Historic Preservation (see Historic Buildings and Facilities)

Historic Properties (see Historic Buildings and Facilities)

Qualified Historic Building or Facility
Application and Scoping
47, 52, 55, 57, 61, 69

Holding Cells (see Cells)

Hospitals (see Medical Care Facilities)

Hotels (see Transient Lodging)

Housing at a Place of Education
Application and Scoping
11, 29, 82

Housing Cells (see Cells)

Index and List of Figures

I

ICC/IBC
Application and Scoping
42

Intercity Rail
Application and Scoping
61, 62, 76

Intersections
Application and Scoping
13

J

Jails (see Detention and Correctional Facilities)

Judges' Benches
Application and Scoping
57, 63

Technical
212

Judicial Facilities
Application and Scoping
62, 89

K

Kitchens and Kitchenettes
Application and Scoping
11, 29, 46, 55, 69, 93

Technical
206-209, 211, 213

Knee and Toe Clearance (see also Toe Clearance)
Technical
106-110, 159, 167, 170, 171, 208, 209, 219, 234

L

Landlord/Tenant
Application and Scoping
23

Lavatories
Application and Scoping
37, 70

Technical
160, 170, 171

Light Rail
Application and Scoping
46, 76

Limited-Use/Limited-Application Elevators
Technical
143

Lines of Sight
Application and Scoping
79

Technical
203-205

Lockers (see also Storage)
Application and Scoping
85

Locker Rooms (see Dressing, Fitting, and Locker Rooms)

Long-Term Care Facilities
Application and Scoping
14, 47, 48, 68, 81, 82

Technical
180, 209

M

Machinery Spaces
Application and Scoping
53

Mail Boxes
Application and Scoping
46, 88

Technical
221

Department of Justice   2010 Standards: Title II and III- 261

Index and List of Figures

March 15, 2011
Application and Scoping
9

March 15, 2012
Application and Scoping
8, 10, 26, 27

Maximum Extent Feasible
Application and Scoping
6, 7, 14, 21, 22, 26, 50, 51

Means of Egress
Application and Scoping
44, 46, 64, 68, 72

Technical
127, 212, 227, 228

Medical Care Facilities
Application and Scoping
13, 30, 48, 81, 90

Technical
180, 186

Mezzanine
Application and Scoping
46, 48, 56-58, 80

Technical
217

Miniature Golf Facilities
Application and Scoping
60, 98

Technical
236

Mirrors
Application and Scoping
70

Technical
160

Motels (see Transient Lodging)

Multi-bedroom Housing Units
Application and Scoping
11, 29

Multi-Story Buildings and Facilities
Application and Scoping
56, 57

N

NFPA
Application and Scoping
43

Technical
186, 213

National Historic Preservation Act (see Historic Buildings and Facilities)

New Construction
Application and Scoping
6, 9, 10, 13, 19, 26, 27, 50, 51, 63, 82, 83, 92

Normal Maintenance
Application and Scoping
21, 44

O

Openings
Application and Scoping
88

Technical
105, 119, 140, 218, 226, 227, 229, 231, 236

Operable Parts
Application and Scoping
55

Technical
115, 116, 125, 146, 147, 159, 185, 195, 199, 209, 218

Index and List of Figures

P

Parking
Application and Scoping
7-9, 23, 24, 48, 55, 61, 65-68, 72

Technical
132, 149-152

Passenger Loading Zones
Application and Scoping
55, 65, 67, 68

Technical
152

Passing Spaces
Technical
119

Path of Travel
Application and Scoping
6-9, 22-25, 51

Percentages, Calculation of
Application and Scoping
38

Performance Areas
Application and Scoping
12, 30, 58, 63, 80

Technical
203

Place of Public Accommodation
Application and Scoping
20-22, 25, 26, 47

Places of Lodging
Application and Scoping
28, 82

Platform Lifts
Application and Scoping
40, 41, 45, 63, 64

Technical
117, 147, 212

Play Areas
Application and Scoping
54, 60, 64, 98, 99

Technical
237-239

Player Seating (see Team and Player Seating)

Plumbing Fixtures
Application and Scoping
70

Technical
165, 171

Pools (see Swimming Pools or Wading Pools)

Post-Mounted Objects
Technical
112

Press Boxes
Application and Scoping
58

Primary Function Areas
Application and Scoping
7-9, 22, 23, 25, 50, 51

Prisons (see Detention and Correctional Facilities)

Private Residences (see Commercial Facilities Located in Private Residences)

Private Residence Elevators
Technical
145

Professional Office of a Health Care Provider
Application and Scoping
20, 21, 25, 56

Index and List of Figures

Protruding Objects
Application and Scoping
54

Technical
111, 113

Psychiatric Facilities (see Medical Care Facilities)

Public Address Systems
Application and Scoping
76

Technical
218

Public Entrance (see Entrances)

Public Use
Application and Scoping
12, 29, 42, 45, 47, 54, 56, 63, 68, 71, 75

Technical
160, 163, 171, 172, 177, 197, 201, 216

Public Transportation
Application and Scoping
21, 25, 55, 56, 67

Putting Greens
Application and Scoping
60, 98

Technical
235

Q

Queues (see also Waiting Lines)
Application and Scoping
74, 87, 88

R

Rail Platforms
Technical
216

Raised Areas
Application and Scoping
53

Ramp(s)
Application and Scoping
8, 23, 24, 45, 47, 63, 67, 100

Technical
106, 112, 117, 127-130, 154, 156, 212, 224, 225, 236, 239, 241

Ramp, Pedestrian
Application and Scoping
8, 23

Rapid Rail
Application and Scoping
76

Reach Ranges
Technical
113, 116, 134, 140, 161, 168, 206, 218, 237, 242

Recreation Facilities (see Amusement Rides, Boating Facilities, Exercise Machines, Fishing Piers and Platforms, Golf, Miniature Golf, Saunas and Steam Rooms, Shooting Facilities, Swimming Pools, Wading Pools, or Spas)

Recreational Boating Facilities (see Boating Facilities)

Referenced Standards
Application and Scoping
40-44, 48, 64, 72

Technical
127, 133, 143, 145, 147, 186, 213, 218, 234, 239

Rehabilitation (of buildings and facilities)
Application and Scoping
21, 44

Index and List of Figures

Rehabilitation Facilities
(see Medical Care Facilities)

Remodeling
Application and Scoping
21, 23, 44

Renovation
Application and Scoping
21, 44

Rental Establishments (see Sales or Rental Establishments)

Residential Dwelling Unit
Application and Scoping
11, 13, 28, 29, 47, 48, 51, 53, 56, 62-64, 66, 67, 71, 72, 88, 91-94

Technical
145, 162, 163, 171, 172, 177, 178, 180, 201, 208, 212-214

Residential Facilities
Application and Scoping
11-13, 28, 29, 53, 56, 66, 67, 71, 72, 88, 91

Restaurants and Cafeterias
Application and Scoping
58

Restricted Entrance (see Entrances)

Roll-in Showers (see Showers)

Running Slope
Application and Scoping
45, 47

Technical
117, 119, 127, 128, 133, 237, 239

S

Safe Harbor
Application and Scoping
8

Sales and Service
Application and Scoping
87

Technical
220, 221

Sales or Rental Establishments
Application and Scoping
20, 21, 25, 26

Saunas and Steam Rooms
Application and Scoping
102

Technical
185

Scope of Coverage
Application and Scoping
11, 27

Seats, Bathub and Shower
Technical
172, 174, 175, 181-183

Section 35.151 of 28 CFR part 35
Application and Scoping
6

Security Barriers
Application and Scoping
64

Self-Service Storage (Facilities)
Application and Scoping
47, 62, 86

September 15, 2010
Application and Scoping
9, 10, 26, 27

Series of Smaller Alterations
Application and Scoping
9, 24

Service Entrance (see Entrances)

Index and List of Figures

Shelters
Application and Scoping
11, 28

Shelves
Application and Scoping
70, 76, 81, 86, 88

Technical
161, 162, 168, 206, 222

Shooting Facilities
Application and Scoping
103

Technical
252

Shopping Center or Shopping Mall
Application and Scoping
20, 21, 25, 26, 56

Showers
Application and Scoping
11, 29, 62, 70, 83, 84

Technical
174, 175-182, 210, 211

Roll-in Showers
Application and Scoping
11, 29, 84

Technical
175-180, 183, 210

Shower Compartments
Technical
174-182, 210

Transfer-type Showers
Application and Scoping
11, 29
Technical
174, 177, 178, 180,183

Signs
Application and Scoping
71-74

Signs (cont'd)
Technical
136, 143, 151, 186, 189, 190, 216, 217, 218

Sinks
Application and Scoping
37, 55, 69

Technical
170, 171, 209

Site Arrival Points
Application and Scoping
55

Sleeping Rooms
Application and Scoping
11, 28, 29, 81, 82

Technical
209, 210

Slope (see Running Slope or Cross Slope)

Soft Contained Play Structures (see Play Areas)

Social Service Center Establishments
Application and Scoping
11, 28

Spas
Application and Scoping
102, 103

Technical
125, 242, 244

Stadiums, Arenas, Grandstands
Application and Scoping
12, 29, 45, 78

Stadium-Style Movie Theaters
Application and Scoping
12, 30

Stairs and Escalators in Existing Buildings
Application and Scoping
57

Stairways
Application and Scoping
45, 68, 72

Technical
139, 153

Start of Physical Construction or Alterations
Application and Scoping
26, 27

Storage (see also Self-Storage)
Application and Scoping
7, 9, 22, 24, 47, 48, 53, 62, 85, 86

Technical
209, 218, 227, 228

Streets, Roads, Highways
Application and Scoping
7, 13, 23, 55

Technical
131, 215

Structural Impracticability
Application and Scoping
6, 19

Structural Parts or Elements
Application and Scoping
21, 44

Swimming Pools
Application and Scoping
102

Technical
242

T

Team or Player Seating
Application and Scoping
64, 79, 80

Technically Infeasible
Application and Scoping
14, 48, 50, 69, 81, 93

Teeing Ground(s)
Application and Scoping
48, 60, 97, 98

Technical
235

Telephones
Application and Scoping
7-9, 22-24, 45, 48, 51, 73-76, 90

Technical
146, 193-196, 199, 201, 211, 212, 222

Temporary Facilities
Application and Scoping
11, 28, 50

Tenant Spaces
Application and Scoping
23, 62

Terminal
Application and Scoping
21, 25, 56, 76

Thresholds
Technical
124, 127, 140, 180

Toe Clearance (see also Knee and Toe Clearance)
Technical
106-110, 159, 167, 170, 171, 208, 209, 219, 234

Index and List of Figures

Toilet Compartments
Application and Scoping
37, 70

Technical
161, 165, 166, 168, 169

Toilet Facilities
Application and Scoping
52, 69

Technical
181, 211, 213

Tolerances, Construction and Manufacturing
Application and Scoping
38

Townhouses
Application and Scoping
12, 29

Transient Lodging
Application and Scoping
11, 12, 28, 29, 47, 48, 56, 62, 64, 71, 82-84

Technical
178, 180, 210

Transportation Facilities
Application and Scoping
51, 61, 76

Technical
214

TTY
Application and Scoping
8, 24, 48, 73-76

Technical
193, 196, 201, 211

Turning Space
Application and Scoping
11, 29

Technical
106, 107, 109, 138, 160, 162, 185, 206, 209, 211-213, 224, 235, 238, 239, 241, 252

Two-Way Communication Systems
Application and Scoping
88

Technical
142, 146, 201

U

Uniform Federal Accessibility Standards (UFAS)
Application and Scoping
8, 9, 10

Unisex Toilet Room
Application and Scoping
69, 70

Urinals
Application and Scoping
70

Technical
170

V

Van Parking Spaces
Application and Scoping
66, 67

Technical
149-151

Vending Machines
Application and Scoping
88

Index and List of Figures

Vertical Clearance
Application and Scoping
54

Technical
113, 151, 152, 196, 237

Vertical Viewing Angle
Application and Scoping
12, 30

Visiting Areas
Application and Scoping
89-91

W

Wading Pools
Application and Scoping
102

Technical
242, 246

Waiting Lines (see also Queues)
Application and Scoping
74, 87, 88

Walk
Application and Scoping
7, 13, 23, 45, 46, 49, 61

Technical
131

Walking Surfaces
Application and Scoping
46-48

Technical
117, 119, 154, 155, 196, 220, 224

Washing Machines
Application and Scoping
70

Technical
115, 185

Water Closets
Application and Scoping
37, 70

Technical
161-169, 182, 210, 211 202-206,
212, 224-226

Water Slides
Application and Scoping
45, 54

Weather Shelters
Application and Scoping
60, 98

Technical
235, 236

Wheelchair Space(s)
Application and Scoping
12, 29, 30, 49, 57, 60, 63,
78, 79, 80, 95

Windows
Application and Scoping
7, 23, 73, 88

Technical
222

Work Surfaces
Application and Scoping
11, 29, 37, 86, 87

Technical
208, 209, 219

Index and List of Figures

LIST OF FIGURES FOR THE 2010 STANDARDS

Figure	Description	Page
Figure 104	Graphic Convention for Figures	39
Figure 302.2	Carpet Pile Height	104
Figure 302.3	Elongated Openings in Floor or Ground Surfaces	105
Figure 303.2	Vertical Change in Level	105
Figure 303.3	Beveled Change in Level	106
Figure 304.3.2	T-Shaped Turning Space	107
Figure 305.3	Clear Floor or Ground Space	108
Figure 305.5	Position of Clear Floor or Ground Space	108
Figure 305.7.1	Maneuvering Clearance in an Alcove, Forward Approach	109
Figure 305.7.2	Maneuvering Clearance in an Alcove, Parallel Approach	109
Figure 306.2	Toe Clearance	110
Figure 306.3	Knee Clearance	111
Figure 307.2	Limits of Protruding Objects	112
Figure 307.3	Post-Mounted Protruding Objects	112
Figure 307.4	Vertical Clearance	113
Figure 308.2.1	Unobstructed Forward Reach	114
Figure 308.2.2	Obstructed High Forward Reach	114
Figure 308.3.1	Unobstructed Side Reach	115
Figure 308.3.2	Obstructed High Side Reach	116
Figure 403.5.1	Clear Width of an Accessible Route	118
Figure 403.5.2	Clear Width at Turn	118
Figure 404.2.3	Clear Width of Doorways	120
Figure 404.2.4.1	Maneuvering Clearances at Manual Swinging Doors and Gates	121

Index and List of Figures

Figure 404.2.4.2	Maneuvering Clearances at Doorways without Doors, Sliding Doors, Gates, and Folding Doors	123
Figure 404.2.4.3	Maneuvering Clearances at Recessed Doors and Gates	124
Figure 404.2.6	Doors in Series and Gates in Series	125
Figure 405.7	Ramp Landings	129
Figure 405.9.1	Extended Floor or Ground Surface Edge Protection	130
Figure 405.9.2	Curb or Barrier Edge Protection	130
Figure 406.2	Counter Slope of Surfaces Adjacent to Curb Ramps	131
Figure 406.3	Sides of Curb Ramps	131
Figure 406.4	Landings at the Top of Curb Ramps	132
Figure 406.6	Diagonal or Corner Type Curb Ramps	132
Figure 406.7	Islands in Crossings	133
Figure 407.2.2.2	Visible Hall Signals	135
Figure 407.2.3.1	Floor Designations on Jambs of Elevator Hoistway Entrances	136
Figure 407.2.3.2	Car Designations on Jambs of Destination-Oriented Elevator Hoistway Entrances	136
Figure 407.4.1	Elevator Car Dimensions	138
Figure 408.4.1	Limited-Use/Limited-Application (LULA) Elevator Car Dimensions	144
Figure 409.4.6.2	Location of Private Residence Elevator Control Panel	146
Figure 410.6	Platform Lift Doors and Gates	148
Figure 502.2	Vehicle Parking Spaces	149
Figure 502.3	Parking Space Access Aisle	150
Figure 503.3	Passenger Loading Zone Access Aisle	152
Figure 504.5	Stair Nosings	153
Figure 505.4	Handrail Height	154

Index and List of Figures

Figure 505.5	Handrail Clearance	155
Figure 505.6	Horizontal Projections Below Gripping Surface	155
Figure 505.7.2	Handrail Non-Circular Cross Section	156
Figure 505.10.1	Top and Bottom Handrail Extension at Ramps	157
Figure 505.10.2	Top Handrail Extension at Stairs	157
Figure 505.10.3	Bottom Handrail Extension at Stairs	158
Figure 602.5	Drinking Fountain Spout Location	159
Figure 604.2	Water Closet Location	161
Figure 604.3.1	Size of Clearance at Water Closets	162
Figure 604.3.2	(Exception) Overlap of Water Closet Clearance in Residential Dwelling Units	162
Figure 604.5.1	Side Wall Grab Bar at Water Closets	163
Figure 604.5.2	Rear Wall Grab Bar at Water Closets	164
Figure 604.7	Dispenser Outlet Location	165
Figure 604.8.1.1	Size of Wheelchair Accessible Toilet Compartment	166
Figure 604.8.1.2	Wheelchair Accessible Toilet Compartment Doors	166
Figure 604.8.1.4	Wheelchair Accessible Toilet Compartment Toe Clearance	167
Figure 604.8.2	Ambulatory Accessible Toilet Compartment	168
Figure 605.2	Height and Depth of Urinals	170
Figure 607.2	Clearance for Bathtubs	172
Figure 607.4.1	Grab Bars for Bathtubs with Permanent Seats	173
Figure 607.4.2	Grab Bars for Bathtubs with Removable In-Tub Seats	173
Figure 607.5	Bathtub Control Location	174
Figure 608.2.1	Transfer Type Shower Compartment Size and Clearance	175
Figure 608.2.2	Standard Roll-In Type Shower Compartment Size and Clearance	176

Index and List of Figures

Figure 608.2.3	Alternate Roll-In Type Shower Compartment Size and Clearance	176
Figure 608.3.1	Grab Bars for Transfer Type Showers	177
Figure 608.3.2	Grab Bars for Standard Roll-In Type Showers	177
Figure 608.3.3	Grab Bars for Alternate Roll-In Type Showers	178
Figure 608.5.1	Transfer Type Shower Compartment Control Location	178
Figure 608.5.2	Standard Roll-In Type Shower Compartment Control Location	179
Figure 608.5.3	Alternate Roll-In Type Shower Compartment Control Location	180
Figure 609.2.2	Grab Bar Non-Circular Cross Section	181
Figure 609.3	Spacing of Grab Bars	182
Figure 610.2	Bathtub Seats	183
Figure 610.3	Extent of Seat	183
Figure 610.3.1	Rectangular Shower Seat	184
Figure 610.3.2	L-Shaped Shower Seat	184
Figure 611.4	Height of Laundry Compartment Opening	185
Figure 703.2.5	Height of Raised Characters	187
Figure 703.3.1	Braille Measurement	188
Figure 703.3.2	Position of Braille	189
Figure 703.4.1	Height of Tactile Characters Above Finish Floor or Ground	189
Figure 703.4.2	Location of Tactile Signs at Doors	190
Figure 703.6.1	Pictogram Field	192
Figure 703.7.2.1	International Symbol of Accessibility	193
Figure 703.7.2.2	International Symbol of TTY	193
Figure 703.7.2.3	Volume Control Telephone	193
Figure 703.7.2.4	International Symbol of Access for Hearing Loss	194

Index and List of Figures

Figure 704.2.1.1	Parallel Approach to Telephone	194
Figure 704.2.1.2	Forward Approach to Telephone	195
Figure 705.1	Size and Spacing of Truncated Domes	197
Figure 707.6.2	Numeric Key Layout	200
Figure 802.1.2	Width of Wheelchair Spaces	202
Figure 802.1.3	Depth of Wheelchair Spaces	203
Figure 802.2.1.1	Lines of Sight Over the Heads of Seated Spectators	204
Figure 802.2.1.2	Lines of Sight Between the Heads of Seated Spectators	204
Figure 802.2.2.1	Lines of Sight Over the Heads of Standing Spectators	205
Figure 802.2.2.2	Lines of Sight Between the Heads of Standing Spectators	205
Figure 804.2.1	Pass-Through Kitchens	207
Figure 804.2.2	U-Shaped Kitchens	208
Figure 810.2.2	Dimensions of Bus Boarding and Alighting Areas	215
Figure 810.3	Bus Shelters	216
Figure 810.10	(Exception) Track Crossings	218
Figure 903.4	Bench Back Support	220
Figure 904.3.2	Check-Out Aisle Counters	221
Figure 904.4	(Exception) Alteration of Sales and Service Counters	222
Figure 1002.4.4.3	Protrusions in Wheelchair Spaces in Amusement Rides	226
Figure 1003.3.1	Boat Slip Clearance	230
Figure 1003.3.1	(Exception 1) Clear Pier Space Reduction at Boat Slips	231
Figure 1003.3.1	(Exception 2) Edge Protection at Boat Slips	231
Figure 1003.3.2	Boarding Pier Clearance	232
Figure 1003.3.2	(Exception 1) Clear Pier Space Reduction at Boarding Piers	233
Figure 1003.3.2	(Exception 2) Edge Protection at Boarding Piers	233

Index and List of Figures

Figure 1005.3.2	Extended Ground or Deck Surface at Fishing Piers and Platforms	235
Figure 1007.3.2	Golf Club Reach Range Area	237
Figure 1008.3.1	Transfer Platforms	240
Figure 1008.3.2	Transfer Steps	241
Figure 1009.2.2	Pool Lift Seat Location	243
Figure 1009.2.3	Clear Deck Space at Pool Lifts	244
Figure 1009.2.4	Pool Lift Seat Height	244
Figure 1009.2.8	Pool Lift Submerged Depth	245
Figure 1009.3.2	Sloped Entry Submerged Depth	246
Figure 1009.3.3	Handrails for Sloped Entry	247
Figure 1009.4.1	Clear Deck Space at Transfer Walls	247
Figure 1009.4.2	Transfer Wall Height	248
Figure 1009.4.3	Depth and Length of Transfer Walls	248
Figure 1009.4.5	Grab Bars for Transfer Walls	249
Figure 1009.5.1	Size of Transfer Platform	249
Figure 1009.5.2	Clear Deck Space at Transfer Platform	250
Figure 1009.5.4	Transfer Steps	250
Figure 1009.5.6	Size of Transfer Steps	251
Figure 1009.5.7	Grab Bars	251

www.ingramcontent.com/pod-product-compliance
Lightning Source LLC
Chambersburg PA
CBHW031828170526
45157CB00001B/218